# The Paradox of Paradise
## A Second Home Adventure

Richard A. Geudtner

Copyright © 1996 Richard A. Geudtner

First Edition

All rights reserved. Reproduction in whole or in part of any portion in any form without permission of publisher is prohibited.

Amherst Press
A Division of Palmer Publications, Inc.
318 N. Main Street
Amherst, Wisconsin 54406

Library of Congress Catalog Number 96-095022

ISBN 0-942495-59-4

## Dedication

To Jan who shared the dream and participated in its creation.
Through her love and effort the house became a home.

*and*

To Muggs with whom I shared its creature comforts,
and the wonders of its natural world each day of her life.

## Acknowledgments

The author deeply appreciates the comprehensive
editorial reviews given the draft manuscript
by Jim Houy and Betty Parsons and the
thorough proofing provided by Ann Bernard.

## Illustrations

In pen and ink by John Dioszegi
of Highland Park, Illinois,
an artist by profession
and a friend and companion by avocation.

# Contents

Preface . . . . . . . . . . . . . . . . . . . . . . . . . . . . . . . . . . . . . ix

1  The Way It Was . . . . . . . . . . . . . . . . . . . . . . . . . . . . 1
2  The Search . . . . . . . . . . . . . . . . . . . . . . . . . . . . . . . . 9
3  Frustration . . . . . . . . . . . . . . . . . . . . . . . . . . . . . . . 15
4  Design Dilemmas . . . . . . . . . . . . . . . . . . . . . . . . . . 21
5  Getting Underway . . . . . . . . . . . . . . . . . . . . . . . . . 31
6  Our Participation . . . . . . . . . . . . . . . . . . . . . . . . . 36
7  Tornado . . . . . . . . . . . . . . . . . . . . . . . . . . . . . . . . . 43
8  Taking Shape . . . . . . . . . . . . . . . . . . . . . . . . . . . . 50
9  Move-In . . . . . . . . . . . . . . . . . . . . . . . . . . . . . . . . 56
10 The First Fall . . . . . . . . . . . . . . . . . . . . . . . . . . . . 63
11 The Storm/Armoring The Shore . . . . . . . . . . . . . . 69
12 Necessary Arrangements . . . . . . . . . . . . . . . . . . . 76
13 Winter . . . . . . . . . . . . . . . . . . . . . . . . . . . . . . . . . 83
14 Spring . . . . . . . . . . . . . . . . . . . . . . . . . . . . . . . . . 95
15 High Water . . . . . . . . . . . . . . . . . . . . . . . . . . . . 103
16 Planning Ahead . . . . . . . . . . . . . . . . . . . . . . . . . 109
17 Trees . . . . . . . . . . . . . . . . . . . . . . . . . . . . . . . . . 118
18 Summer . . . . . . . . . . . . . . . . . . . . . . . . . . . . . . . 130
19 The Overlook . . . . . . . . . . . . . . . . . . . . . . . . . . . 144
20 The Garage . . . . . . . . . . . . . . . . . . . . . . . . . . . . 150
21 The Access Lane . . . . . . . . . . . . . . . . . . . . . . . . 160
22 Holding Tanks . . . . . . . . . . . . . . . . . . . . . . . . . . 167
23 Conservancy . . . . . . . . . . . . . . . . . . . . . . . . . . . 177
24 Autumn . . . . . . . . . . . . . . . . . . . . . . . . . . . . . . . 199
25 Coming Home . . . . . . . . . . . . . . . . . . . . . . . . . . 214

# Preface

While growing up, there always seemed to be a friend or two who, when asked what they were doing over the weekend, would inevitably reply, "We're going to our summer home." I remember the neighbors packing the car and loading kids and dogs for the usual weekend "at the lake." It always left within me feelings of childhood envy, wishing it was my family. I wistfully conjured up dreams of the north woods and other romantic notions of what it must be like to enjoy all those things such as boats and fishing, the forest and the lakeside vistas which were denied us ordinary kids. We who may only have had a glimpse of such things in passing, or on a week's vacation. Or worse, living it totally vicariously with only one's imagination stimulated by a friend's tales told on Mondays, when he or she had reappeared to spend the ho-hum summer weekdays awaiting the next Friday night departure.

No matter that the summer home in the north woods might be only a hand-me-down cottage on a lake in northern Illinois ringed with cottages more closely spaced than their in-city dwellings may have been. That the lake was overrun with overpowered outboards and waterskiers, making a venture upon its waters to swim or fish or sail a perilous undertaking. Or that the "forests" gave way to cornfields at the rear lot lines of the lakeside property. To me left behind, it would yet have been nirvana.

To those lucky souls, both grown-ups and children, who make the weekend treks and fight the traffic, the vacation retreat does indeed satisfy the spirit's need for a break from routine and the uplift received from experiencing a more tranquil world of the out-of-doors. Possibly the only dissent may be from the wives and mothers upon whose shoulders always seem to rest the unending responsibilities of washing the prior weekend's clothes and linens and preparing for the next onslaught of lakeside tranquility. Tranquility, which always seems to include the inevitable guests and relatives, sometimes invited by the

## The Paradox of Paradise

head of household without consultation—and others showing up unannounced, coveniently just prior to lunch. Even these wives and mothers must feel the balm of the early morning when the air is still and the lake is calm. While having coffee on the porch, her only companions are the avian kind twittering away in the foliage above and the fisherman in the distance silently casting the shallows, savoring his version of nirvana prior to the mechanized storm.

This second home phenomena seems to be present throughout the modern world where wealth (and I am speaking relatively) and material amenities and transport of the twentieth century have given ordinary people the leisure time and means to enjoy this Earth's natural world. The "summer" home may be along the ocean shore, on a tropical island, a creek, a small lake, a mountainside or an abandoned strip mine, or just in the rural countryside somewhere. It may be characterized as a cottage, a cabin, a dacha, a farm, a tent or a year-round house, many of which today put to shame our parent's only home. It is not exclusively a North American pursuit. I have been on the highways of Europe on a summer Friday evening when it looked as though everyone was fleeing another world war, carrying along all the baggage, impatience and frustration attendant to desperate flight. The Europeans seem to take flight with a unique vengeance and determination along the autobahns and autostradas of the continent, yet stall with resignation at the frontier checkpoints backed up by the thousands—choking the very forests they love with exhaust.

Once, while being driven from the airport to the town of Nuku'alofa on the island of Tongatapu in the kingdom of Tonga, a million miles from nowhere in the South Pacific Ocean, I was passed by a black Cadillac limousine. As there were only a handful of automobiles on the entire island at that point in history (1966), and they were all rusted "beaters" being held together with bailing wire, my curiosity was piqued at this ostentatious sight along the unpaved road. With great pride and reverence, my driver answered that it was the King being driven to his summer home.

To put this into perspective, the island of Tongatapu is only seventeen miles across at its extreme breadth. So it just illustrates that this "summer cottage," second home thing has its devotees across many of the world's cultures. Given a little leisure time and sufficient wherewith-all to pay the bills, we all seem to want our place in the sun where

Preface

it is possible to shed the tensions of the working world. There are those very practical folks who say they would never want to get saddled with the costs of and enslavement to a second home ("why I'd always feel we had to go there," or, "it would keep us from doing other things," or, "I worry so much about our home when we are away that I wouldn't want to have to worry about a second home"). While spending a weekend as guests in a friend's home at a lake or along the seashore, even these naysayers have experienced the real sense of letting down in a private place away from it all. Also they may have had a nascent feeling that it would be nice to have this piece of the natural world with its views and scents and wild things to return to at will—whenever the heart desired—without having to make a reservation or be invited.

⚓ ⚓ ⚓ ⚓

The latter part of the twentieth century is seeing a transition in second home concept from that of the uninsulated frame cottages of our parents and grandparents to larger year-round homes suitable for enjoyment of every season in temperate climates. The former had to have their plumbing drained and windows shuttered each fall to remain unused for three-quarters of the year. As time passes, many of these new generation second homes are becoming retirement abodes with a complete divorcement of the occupants from urban life. The possible exception may be those well-to-do folks who retain a pied-a-terre in the city to enjoy the sophisticated pleasures of urban life when the spirit moves them—sort of a vacation from a vacation.

It's curious over the span of time how change has come about to these "north woods" settings once virtually abandoned after Labor Day weekend, or at least as the leaves started to fly. A drive down the rural byways at night on a winter weekend in past years revealed total darkness within the snow-covered landscape. Today more and more lights are on, evidence of winter use. A realization that the off-seasons each provide their own joy of the out-of-doors: the color changes in autumn, and walks in the woods on brisk days; the exhilaration of skiing the silent paths of the wind-buffered forests; to late winter when the days become longer and the sun's rays begin to give warmth instead of simply light, leading to spring's promise of the first wildflower and the perceptible thickening of the forest canopy. Dramatically, the breakup of ice on the lakes overnight transforms the silent grip of winter into the blue open water promise of nature's new year.

Resorts and restaurants that used to be open only during the summer season are remaining open to satisfy the inclination of people to experience the out-of-doors off-season without the frenetic pace accompanying the summer season. These off-season pleasures are less contrived, all brought into focus by experiencing the simple beauty of natural surroundings undisturbed by summer's distractions.

As desirable property becomes less and less available, lakeshore and oceanside property values have shot into the stratosphere. With the cost of property so dear, prospective second home owners buy ever smaller lots, on which they desire ever larger homes with more and more amenities. Now the latest phenomena to hit the scene is the purchase of older cottage properties only for their land value. The purchasers raze the rotting clapboard summer cottages and replace them with today's version of a second home.

Communities that have lax or nonexisting ordinances controlling quality of development find the natural features which attracted the initial settlement being destroyed. The Johnny-come-latelys are building upon marginal properties without any sensitivity to the natural surroundings. I cannot understand the logic in buying a small piece of forested land only to cut down all the trees to build the large home. And then, "relandscape" with a few scrawny specimens which will never in the remaining life of the property owner amount to anything.

The irony is we are all guilty to some degree or another of destroying what we originally sought. A quiet place to enjoy the beauty of nature as a restorative to the spiritual vacuum of our competitively driven, technologically dominated world.

A friend of mine searched long for property on a totally undeveloped lake. Find it he did, on a beautiful and pristine lake in the Upper Peninsula of Michigan; one of the few properties around the lakeshore not held by a public corporation. Sweat, blood, tears, toil, and not a little money was spent creating his dream—a beautiful log home nestled into its natural setting, undisturbed until twelve years later when surveyors began plating the lots of a second home subdivision around the lake. The land had been sold to a paper company which had decided to cash in on the second home craze, and score it did! Within two years, all except two lots were sold and cottages were springing up in the lakeside subdivision. Each new owner enthusiastically captured his slice of the wilderness, only to be rudely awakened one morning to the sound of chain saws and bulldozers of contract loggers beginning the clear-cut of all the surrounding forest for pulpwood, leaving a blighted, desolate landscape their legacy for several future generations.

So who cares? The second home phenomena is self-evident, and the results are predictable, if sad, wherever. It does hint at the ethos of the writer and is a prelude to the story of this person's involvement with a second home. How could someone so avowedly uninterested in the subject and conscious of its negatives become engaged in an architect's "busman's holiday" (I had sworn never to design or build anything for myself because I could not walk away from it at night), and then be swept into a far-reaching endeavor which became nothing less than a raison d'etre.

From a passive interest piqued during occasional weekends spent at friends' homes, my stimulus always sank back to dormancy with time, other involvements and the rationalization of the aforementioned negatives. Once involved with the project, I experienced an evolutionary transformation to an unequivocal love for a place. Along the way, these feelings have been accompanied not only with an unflagging interest and commitment, but also a continual spiritual uplift and joy with my good fortune of being there. Yet this inspiration has been tempered with unenvisioned hardships and tragedy caused by the vicissitudes of nature and human nature. If there is a difference in my enthusiasm from what might be others, it has been because of "the place." I became devoted to a place and a creation (my creation for whatever it's worth)—not to the concept of second home ownership itself.

# Chapter 1
# The Way It Was

Sailing boats has been the writer's lifelong avocation and downright passion. From the tender age of ten when I ran off from my parents while sightseeing on Mackinac Island because I was determined to see the three-masted schooner in the harbor, the pastime of sailing has been a consuming and unidirectional force throughout life's course.

When a teenager, my dedication to the sport afforded access to wonderful racing crew berths aboard winners. It gave me impetus toward a Navy scholarship and a commission in that branch of the armed forces. It drew me hitchhiking home from the university on weekends to sail. It stimulated me to learn navigation. During the years "sailing" the oceans on a destroyer escort, it coalesced within my mind a recognition that if ever I was to have the wherewithal to realize my always present dream of owning and sailing my own yacht, I had to remove myself as quickly as possible from the stultifying penury of military life. I was compelled to seek my fortune, limited only by my own shortcomings and my willingness to devote myself to this purpose—with regular lapses for sailing, of course.

Upon leaving the service, my brother and I opened our architectural firm in suburban Chicago. He with eleven years experience in the field and I with the highly useful experience to the building design world of how to drive a ship.

On the other hand, there is no substitute for determination and persistence in selling architectural services as in doing anything. Selling in the daytime, drawing at night and on weekends paid off. Yet, summer weekends, even then, always managed to be reserved for sailing. Devotion to sailing and workaholic habits are often a deadly combination when pitted against marriage. My father always said sailing boats and hanging around yacht clubs was a sure path to the divorce court. I didn't disappoint him in this simplistic, but prescient, admonition. In 1973 I found THE yacht, a 48-foot yawl, and convinced my brother to join me in this endeavor with me being captain. As the reader might imagine, this joint venture in yacht ownership lasted less than two years when he decided that five days a week with some nights and weekends thrown in was enough hounding. Better the sporting venture down the tubes than the business partnership. Thus, it became sell the boat or buy him out, and I certainly was not going to bow to failure having once realized my goal. With a knot in my stomach swallowing the expense (on top of a recent divorce), the *Aurora* became mine alone to cherish in the season and agonize over in the winter.

With the boat spending its winter seasons laid up at a shipyard in Sturgeon Bay, Wisconsin, I began to spend what may seem to many an inordinate amount of time at the yard decommissioning the vessel, overseeing the yard's winter work of painting and varnishing, rebuilding and improving. This required in my scheme of things bi-weekly inspection trips, design sketches, and in-progress decision making, along with the inevitable monthly discussions of invoice justification. Springtime meant, and still means, the weekends of work with the crew recommissioning, the sea trial and the sail down Lake Michigan to the summer mooring in Waukegan, Illinois. Attendant to all of this time in Sturgeon Bay was the necessity of lodging, which in the middle of winter in years past was not easy to find. Although Door County had literally dozens of motels, hotels, lodges and resorts within twenty or thirty miles of Sturgeon Bay, almost all were "Closed for the season."

The irony is that it was probably as difficult to find a place to stay on a summer weekend with the season in full swing. No Vacancy signs are routinely posted to discourage those unfortunates without reservations. "Full swing" is a misleading description as most folks coming to the Door Peninsula do so for its scenic beauty, outdoor recreation and a summer climate remarkably tempered by the cold waters of Lake Michigan and Green Bay flanking the narrow peninsula.

The Stardust Motel was a lucky find–lucky for me and the owners struggling through their winter's dry spell of an occasional business traveler at the shipyards. Many were the times all eleven units were dark except for my door. And damn, it was cold in the room, because no matter that I gave the proprietors fair warning of my arrival, they always seemed to forget to turn up the heat. It takes forever for electric baseboard radiation to bring a space up to reasonable living temperature when it is ten degrees outside and a thirty-mile-an-hour wind is driving up the ice-covered bay, by the shore of which stood the Stardust.

Convenience and not much choice made the "Acropolis" Cafe a block up from the waterfront the place to eat breakfast. Acropolis was euphemistically derived by my crew from the name on the tattered awning. Greasy eggs and bacon were pounded down in my enthusiasm to see to *Aurora's* needs and inspect the latest project being undertaken by Palmer Johnson's peerless carpenters and painters.

Friendships blossomed at the yard despite the monthly joust over hours billed. Often the end of the day resulted in an invitation to have supper with the yard superintendent at his home in the country. The dining room overlooked a frozen pond and snow glistening into the gloom of the winter forest beyond. This has always been a treat–the hearty country fare, and inevitably, newly made chocolate chip cookies. It was almost as though Joyce knew I was coming. Jim and Joyce plant and tend a vegetable garden the size of a small field. During fall, visits always resulted in take-home treats of tomatoes, squash and other garden-fresh booty, not even mentioning the jars of raspberry, strawberry and rhubarb jam which to this day are still coming my way.

\* \* \* \*

Door County is the sort of place that is ideal for ambling about the rural byways by car, or even bicycling. It is, at most, a twenty-minute leisurely drive across from shore-to-shore at its broadest. The countryside forests of pine and cedar, birch and maple are interspersed with dairy farms and hardscrabble fields. In years past, the fences were all rock walls picked by hand from the shallow till. Scratch the surface and bedrock's glaciated crust is exposed. Often it is the surface feature, bearded only with lichen and juniper, tough farming indeed for those Scandinavians who first scratched a living off this land. From the rocky bluffs overlooking Green Bay across the bedrock spine, its fields filled with daisies during the season, to Lake Michigan's coast of sandy beaches and sea-eroded limestone buttresses shelving under the water's surface, the natural vistas have always beckoned me to explore these pathways. To enjoy time and again the same scenes of forest and lakeshore.

My drives also opened the floodgates of memory, vacationing in these spots with my parents during childhood. Whether alone or with my new family, it was with great anticipation I would look forward to finishing work on *Aurora* and having a few hours to meander along my favorite routes.

Although for years these meanderings were just that, pure aesthetic enjoyment, the Stardust was getting to be increasingly old hat. With the burgeoning sportfishing for salmon and trout, a winter's night at the Stardust was one of blessed tranquility, albeit cold, compared to the boisterousness of beery fishermen in spring and fall. Typically, they went to sleep at midnight only to be slamming about their rooms well before dawn, venturing forth for a day's hungover fishing. Everything being clearly audible through the resonant walls, efforts at sleeping became a torturous impossibility during the shank of the night. It was then it must have occurred to me that having one's own place might not be a bad idea, anticipating continued sojourns to Sturgeon Bay on *Aurora's* behalf into what seemed the forever future. As getting tired of or losing interest in my mistress was an unlikely prospect. From a practical standpoint, staying in a motel was soon to become difficult, if not precluded, because we acquired a puppy which I did not want to leave behind on my frequent journeys to Sturgeon Bay. Still, the previously mentioned negatives raised their chorus of gloom to emphasize the adverse aspects of having a second home. Building a home to satisfy this incipient inclination at this point had never even been given a moment's consideration.

The Way It Was

The seed was planted within my subconscious as subtly as one might be from a tree drifting to the earth amongst thousands of others, randomly germinating where myriad of its kind lay dormant. My meanderings about became somewhat more directional, and my awareness of houses and For Sale signs became more acute. Years before I had discovered an unusually scenic route winding along the Lake Michigan littoral not far north of Sturgeon Bay. The narrow road serpentined along for seven miles beneath a canopy of forest, backdropped by the vista of sky and lake through the trees with only an occasional home to intrude into the natural scene. No condominium developments, no resorts, no farms nor gift shops to mar the sylvan beauty of the drive.

My interest in the Door Peninsula had always been upon the Lake Michigan side of the thumb. The Green Bay shoreline has its rugged cliffs, quaint villages and natural harbors, and has been, since original settlement, the area of greater population density.

For several generations this has been the location of the primary resort activity, leaving the east shore with its cooler, damper climate, its shelving rocky shores relatively unapproachable by boat, and its cedar forests and swamps commercially undeveloped. Each side of the Peninsula has its devotees and passionate defenders, the louder and more numerous voices resounding from the Green Bay shoreline communities. That's fine with me, being one drawn to the solitude of the natural world. To tuck myself as far away as possible from the traffic congestion, the gift shops, golf courses, campgrounds, restaurants and neighbors had been a dream for as long as I recollect. Whenever my fantasies of a second home, fleeting though they were, took possession of my otherwise sensible mind, I envisioned a home in the forest (my forest) fronting on a lakeshore of grand proportion—any monarch's dream!

In the real world of Door County real estate in 1983, there were only a handful of houses on the market along my favorite drive. A few I casually walked around only to be turned off by the appearance, or size, or price, if I bothered to call the realtor. Either they were in the uninsulated chicken coop category with sagging linoleum floors and grotesque lighting, or were overpriced five-bedroom pieces of junk. All of them were suffering from the summer place furnishings syndrome of, "Don't throw it out, take it to the cottage." One log cabin on a beautiful, but small, duned lot, struck my fancy until I poked around with my rigging knife and almost lost the blade in rotten logs and window frames. The problems were repairable, but it was under contract— in retrospect a blessing of unavailability. Recollecting that I had previously seen the For Sale sign six months past, had written the phone number down and never did anything about it, my casual attitude reckoned it wasn't meant to be, and went back to thinking about *Aurora*. As appealing as was the property with its small log house snuggled within the embrace of the cedar trees, as grand as was the view of the lakeshore through the bay windows, the 60-foot-wide lot with neighboring houses close aboard just did not have the necessary attributes to have seeded and nurtured that love of place which has swept me off my feet and kept me lastingly breathless. Good fortune blessed me again as it did in my search for and eventual discovery of the beautiful yawl in Ft. Lauderdale, Florida, which became *Aurora*; a serendipitous happening of great and lasting consequence.

My random examination of houses for sale indicated by the realtors' signs continued casually and critically. Critically because of my picky attitude and because of my profession. I see many things wrong with most any house I have ever looked at. In truth these houses were artless, with little benefit of any professional planning of layout and aesthetic design. Hearing my repetitive criticism, my wife Jan commented that if I ever was going to undertake this venture in second home ownership, the only way I would ever be satisfied would be to find a parcel of vacant land and design and build my own house. Her opinion was I could look until kingdom come without finding anything with which I would not find fault.

This radical advice was received with a reflexive shrug of the shoulders, considering my often stated conviction never to build anything for myself because I could not walk away from it at night. A personalization of all the frustration seen on a daily basis with clients and contractors; requirements exceeding budgets, and differences in perception and taste between husbands and wives. Contractors' bids exceeding estimates. Reductions in scope never meeting expectations of savings, yet additions always exceeding assurances. On-the-job carelessness of contractors and their complete lack of sensitivity to the surrounding natural environment. No-shows, broken promises and prolonged schedules. A lack of attention to the drawings sometimes from laziness, or worse, a determination to do it their way, not as depicted. Contractors' casualness preparing monthly payout documentation, and the architect's fight to get them back to complete the corrective nitty-gritty.

It never fails to amaze me why it is that small contractors, when they realize that they cannot, or will not, show up as promised, have never heard of a phone to say, "Look, I've got a problem, I can't be there at 8, but I can get there at 2:30"... or whenever! I have never run across a single group of people with a trait so common across personality and vocational endeavor. It reminds me of my electrician/plumber/heating contractor in Door County whose answering machine makes the statement, "I'm sorry I'm not here to take your call. I'm either working or playing with my granddaughter. Your call is important to me so if you will leave a message, I'll call you back today... or tomorrow... or for sure the next day. Bye." I have learned not to hold my breath. Once in a while I have reached his wife who has, over the course of time, apparently accommodated herself to these calls by saying that she doesn't have the faintest idea when he will be home. One learns to call before 7 a.m., and call and call repetitively until you luckily squeeze in between every other customer and supplier who's undertaking the same exercise. But, don't brush your teeth in between attempts because you might find he escaped and you've ended up with the damnable machine. To be fair to Roger, he is in his quiet manner a very friendly, sincere person who has responded quickly to emergencies, despite this endemic fault.

## Chapter 2
# The Search

Mulling over Jan's observation, I came eventually to agree that buying land and designing our home would be the only path to realizing this simmering dream. At that moment all prior reason was lost. Having made up my mind this was to be the course, I began a far more organized and dedicated approach to acquiring a piece of vacant property than I had in my casual search for a house. In the seven miles along the drive there were only three lakeshore sites for sale. One was very small and shallow, the second was astronomical in price and difficult to build upon because of the topography. A lakeside dune was too steep on both front and back sides. A rather difficult problem in siting a home unless you are simply willing to remove the dune, consequently interrupting the character of the foreshore, like a bite out of a slice of watermelon; to say nothing of destroying the trees on the dune. The seller lived in California and was in no hurry as she had just gotten her asking price on the adjacent 200 feet. I did not want to pay so much for the land that building the house would have to be postponed while I dug in to recoup the fortune buried in sand. I'd committed, and I was going all out for results—as is my nature. This, of course, breeds frustration when success does not appear quickly.

The third site was identified with a sign barely legible due to the ravages of time and weather and a few helpful bullet holes. Almost obscured by foliage, it advertised an indeterminate size property by an out-of-area realtor. The drive made a sweeping turn inland at this point and a gravel access road forked off through forest and swamp for a quarter of a mile or so before rejoining the paved road. This gravel lane serviced three homes along the way. I had ventured down this byway previously without a second thought. At the fork in the road, a sandy track split off to wind its way through a dense pine and birch forest toward a point of land on the lakeshore several hundred feet down the pathway. Along the way it passed across a small grassy meadow opening onto the beach, only to dive back into the woods until it became indistinct at the point.

We came upon this path in springtime and, walking out onto the beach, gazed upon a stage set by nature. Large white pine and cedar backdropped a waving sea of tall grass undulating to the tree line. To the south a creek rippled through its sandy delta emptying into Lake Michigan. The small meadow was fascinating to me because I had sworn not to cut major trees on any site to facilitate building the house. I sensed a spectacular architectural opportunity being offered by this grassy clearing fronting upon the beach with its wonderful vista southwestward for miles down the forested crescent coastline. There was no vista northeast with the grassy clearing tucked in behind the wooded point, well protected from the bitter northeasterly storms of autumn and winter. It occurred to me then and there, if I was going to build a house I had found the place, and I mentioned this to Jan.

It was impossible to ascertain any boundaries or how many tracts there were, so my enthusiasm was tempered with the possible reality of it not being suitably divisible or buildable, or no longer for sale, without any consideration of price.

A phone call to the out-of-area realtor divulged that it was a series of five contiguous large tracts, realtor owned. I asked if he would send me a survey which arrived several days later. The document was not much help as it did not show the location of the creek or any other benchmark to aid in identifying where the individual lot lines were located along the road.

The next evening we drove north to satisfy our curiosity, our two-and-a-half-month-old Scottie, Muggins, getting her first taste of the trek to Door County. Three potty stops later we were at the Stardust Motel spreading newspapers across the carpeting to prevent an accident on the shag by the dog, who was to remain invisible. The proprietors had given us special dispensation for Muggs in their "No Pets Allowed" accommodation. Everything went well until 2:30 a.m. when the ubiquitous fishermen began banging around next door and Muggins became the noisy invisible dog, reacting to strange voices and doors slamming.

By 9 a.m. we were at the property searching for corner posts. It would have been impossible to drive out the sandy track to the beach because of the encroaching vegetation. In fact, it wasn't easy walking with legions of gargantuan mosquitos attacking every foot of the way until we reached the open beachfront where an onshore breeze held the swarms at bay within the forest. Once out on the beach it was impossible not to have an appreciation (if only momentary) of the sunny June morning on the pristine shore of this inland sea. Momentary, because I had things to do and mosquitos to contend with. It became apparent that the creek split Tract 19. Just behind the beach, 18 contained a slough formed by an oxbow cutoff of the creek in times past. Seventeen appeared to contain the clearing on a sandy knoll, partially shaded by the spreading branches of a large red oak, a rather rare specimen in this northern seaside forest. Tracts 15 and 16 (the northeasterly most) did not seem to have much to offer my purposes as the lakeshore ridge fell off into a swamp reaching inland to the gravel access lane. As it turned out, the swamp tongued into the back half of 17. The exact nature of the topography in a northern forest is not terribly easy to discern flailing about through the dense undergrowth in the buggy month of June.

The mosquitos and the swamp may well have discouraged others but not this potential property owner. I'd found the place and the mosquitos of June were not about to rain on my parade. They would abate with the passage of summer and become no cause for aggravation to inhibit enjoying all the offerings of the other seasons. June is still my least favorite month in residence, though some years are noticeably worse than others depending upon the dryness of late spring. I believed the swamp, though a breeder of mosquitos, to be an asset to my enjoyment because of the variety of bird and animal life supported by this natural feature.

After a picnic lunch at a public access point to Lake Michigan, we stopped at the county building and purchased a County Zoning

Ordinance, a review of which was essential to an understanding of the limitations of use on this property. Upon reaching home I called the out-of-area realtor and said I was interested in buying Tract 17, as I did not believe 18 and 19 were buildable from my analysis of the zoning restrictions. Certainly 19 was not with the creek running through it, and a setback of 75 feet required on either side of any navigable stream. It's hard to fathom why this meandering stream no more than 10 feet wide and clogged with windfalls would be classed as a "navigable stream," but bureaucracy is often inscrutable. The Wisconsin Department of Natural Resources' definition of navigable is, as I understand it now, a watercourse with defined banks upon which one could float a canoe one day a year. Tough to fight that interpretation.

George was a crafty old guy who had been around the real estate game a long time, and he proved it quickly. "I am saving 17 for myself but would be happy to sell you 18 or 19"... at his astronomical price, of course. Not having much experience in this real estate thrust and parry, I felt a little abused that he would think me so stupid to consider a purchase of undevelopable property. Reflecting upon this, I was certain there was no way this elderly man was going to build a house on 17. He was more aware than anyone else that 17 was the prime parcel. If he found a pigeon to take 18 and 19 off his hands, it would only enhance the value of 17. Be that as it may, he wasn't of a mind to make a deal at this point of time, and this was his excuse. I have learned in the course of the last few years of purchasing properties that if a person does not want to sell you may as well forget it, unless you are willing to pay the unrealistic asking price. So just cool it for awhile and hope for the best. At that time it was not easy to bank the fires of desire.

Summer of '84 was frenetic with work and sailing. Time was almost nonexistent to begin looking for other properties which realistically was necessary because George might never change his mind, or he might find that pigeon. As soon as *Aurora* was put to bed for the winter the search began again with the help of a Sturgeon Bay realtor. We identified suitable properties not on the market, and he researched the tax records and contacted the owners. The chance of success of this method is small at best as the inquirer figuratively, if not actually, is appearing at the property owner's door with hat in hand. Owning real estate does things to peoples' psyches, creating visions of grandeur, especially when they knowingly are sitting on property in a limited marketplace. Again three sites were eventually pursued with one owner willing to sell–at top dollar. Ironically, in retrospect, each of these three sites has subsequently been sold at or above the pricing considerations of '84. So who's got the savvy? It appears to me that it's the local folks smugly sitting on their piles of golden rock and sand, as demand and price for lakeshore property ever escalate.

By late October with prospects for success becoming dimmer, our search extended fruitlessly farther and farther up the shoreline away from the locale which had struck my aesthetic fancy. We had simply exhausted all possibilities. With no apparent purpose other than to satisfy some perverse enjoyment, we decided to take a walk along the path through the property which I had months before proclaimed to be THE place. The forest path was clear of the underbrush of dogwood, raspberry, and thimbleberry bushes leaving the last leaves of autumn on the birch and maple trees rustling in the breeze.

Once out upon the beach without the protection of the cedars, the breeze became a stiff onshore wind biting into our cheeks–a foretaste of more bitter things to follow. The noise of the wind and the surf made conversation at any distance impossible without shouting. Remembering the tranquil summer scene along this shoreline, I felt this demonstration of nature's seasonal progression toward winter was being staged to discourage viewers with its bleakness and late October's biting chill. But to me, this cameo performance of wind and sea upon the desolate foreshore could not have been more beautiful in its raw power, nor could it have enervated me more to rededicated action. Let's get back to our friend's house we had borrowed and call George. You just never know.

Within the hour crafty old George was on the phone listening to my rehearsed approach as to why he should reconsider selling Tract 17. I did not get beyond the introduction when he interrupted with, "Well, make me an offer, but I won't sell 17 without 18 and 19." Nearly speechless, I mumbled that was OK with me, and I would respond. He didn't know I wanted all three tracts and would have been terribly disappointed if I could not have acquired the creek and slough. I did have enough presence of mind to comment that I could understand why he would insist that 18 and 19 be included in the purchase of 17.

On the way home Jan described her concept of the house which

## The Paradox of Paradise

coincidentally turned out quite similar to that which I had conceived. Two bedrooms with their own baths, each in a separate wing, separated by a large open living-dining-kitchen area. I sketched this out on an envelope adding in a utility room, closets, and three fireplaces. By the time we reached home, I had noodled and erased until I had everything somewhat in proportion and scale, albeit small.

So excited was I by George's unanticipated response and the prospect of my dream becoming reality, I spent a sleepless night tossing offers through my head, analyzing and reanalyzing their merit and chance of success. In the morning with hollow eyes, an overworked brain and an inconclusive analysis, I finally decided to offer only a token amount for 18 and 19. These were clearly undevelopable, not only in my appraisal, but also in George's awareness as he had inadvertently disclosed over the phone. The offer for 17 would be somewhat below market to afford some give, if necessary. That afternoon I had a real estate sales contract with an accompanying deposit on the way to George via registered mail with a rather short response deadline.

The next three days were an agony of anticipation and dread, until Jan called at work with the news that George had returned the contract signed—no counter proposal, no equivocation—a deal!

# Chapter 3
# Frustration

The offer to purchase was contingent only upon the ability of the soil to test adequately for septic percolation. I quickly arranged to meet a surveyor/civil engineer on-site to undertake two tasks. One, a comprehensive land and topographic survey of the entire property, including locating and identifying the species of all trees six inches in diameter and over. This was to be important reference information for sizing and locating the house to avoid any cutting of significant trees.

The civil engineering half of the one-man team was to arrange for the percolation testing. He was less than optimistic about my chance of success, not because of the absence of good soil (sand being a wonderful percolator) but because of the absence of open ground, except where I wanted to place the house. I was getting my first taste of regulatory generated discouragement. I learned that there were many prohibitions of septic field placements: not on sloping ground, not closer than so many vertical feet from a swamp or seasonally flooded area, not beneath the high-water mark and not in certain characters of sand. What if we can't get the county to issue a septic permit for a normal in-ground field. My engineer said I could always have a pressurized field designed in a mound.

"Where are we going to put a mound of sand 60 by 20 feet?"

"Well, you will have to cut down the trees."

"Oh, for God's sake, I didn't buy a beautiful forest only to strip off all the trees to accommodate a sanitary system." As it happened the only suitable spot was where some of the most beautiful white pines were located.

"There must be another answer."

"Well, you could use a holding tank."

"Fine, how do we go about this? I really would much prefer not to be pumping any waste into the soil."

"Well, the word is that holding tanks are not going to be permitted after the first of the year."

A moment's reflection was all it took to realize that by the time the required percolation testing was done and examined by the sanitarian, we would never have time to get a holding tank system approved in Madison, the state capital.

Instead of plodding along step-by-step, I suggested that we get the sewer contractor out to excavate the trenches for the soil tests. Simultaneously, the engineer could be undertaking the design of the holding tank. Finding out who the test hole excavator was, I called to let him know how I felt about trees and to encourage care with his backhoe. I also asked that he backfill the holes promptly after the sanitarian's inspection. He humored me with assurances that ultimately meant little.

Door County is between the proverbial rock and hardplace when it comes to sanitary waste disposal. Its extremely thin crust of soil over bedrock permits effluents to pass down into the rock without adequate filtration. Very poor or nonexistent septic design standards of the past, combined with the poisons sprayed on the orchards, have created one of the worst groundwater contamination problems in the state. Extremely stringent septic design regulations are obviously necessary and are now enforced. I, in my effluent innocence, thought a holding tank the obvious answer... no in-ground contamination. Ah, but the bureaucracy didn't see it that way because it hadn't developed a moitoring program to ensure that gallonage pumped out of a tank equalled gallonage disposed of in a legal manner. Unscrupulous pumpers dump their loads surreptitiously to avoid payment of the fees, thus aggravating soil contamination. Ergo, better to insist on septic fields than take a chance on a few perfidious pumping contractors—or to undertake more aggressive monitoring of this activity.

Some resorts and condominium developments were having to put in place huge holding tanks which obviously created huge disposal problems, but a single family home is not in the same sewage league.

The thought of having found this appealing place only to be caught on the back side of a possible prohibition of holding tanks, which would preclude my development without ruining the forest, was too much for me to contemplate. So push I did, having a completed survey within two weeks and test trenches dug within a week, but then my efforts ran aground. No leaning on the sanitarian to inspect the holes did any good. He was busy in other endeavors, and if he wasn't, it was raining. The engineer said that it wouldn't do any good to pressure him as the imperative of time (our timing) would not be within his ponderous agenda. It took the sanitarian five weeks to get at his inspection and another to write the report denying a septic permit, thus officially recognizing the categorical need for a holding tank.

When Christmas was upon us and the first of the year only three working days beyond, the engineer informed me he was going to mail the holding tank permit application to Madison that day.

Wait a minute, "How long does it take for the state's review and approval process?"

He replied, "Usually three to four weeks."

My frustration level was rising quickly. "Can't you take it there and walk it through the permit process?"

"Yes, but I don't have any reason to go to Madison other than your permit."

"How long does it take?"

"A day's trip."

"Well, go! I'll pay for your time. It's important." What I didn't tell him was that I wanted to reduce my anxiety level and close the deal which was singularly hinging on this sanitary permit.

His shrug translated into, "It's your money."

On the 29th of December I had the permit and now could close, but George had gone to Texas for a month's vacation.

The septic field test trenches remained open week after week. As these were large trenches (certainly the equivalent of a grave) excavated amidst the large trees, my concern rose with the onset of winter as part of the root systems had been exposed with a consequent loss of grip. If the excavator didn't get at the backfilling prior to a hard freeze, it would either stay that way until spring, or he would be backfilling with frozen chunks of earth. I wanted the sandy soil to be able to settle and compact with rain back into the holes, and be rid of the ugly piles around the site.

Fall and early winter had been unusually warm and dry, but two days before Christmas the weatherman was forecasting the first winter gale with rapidly dropping temperatures, accompanied initially with heavy rain, changing into snow. The worst of all worlds for the trees, an ice and wet snow burden clinging to the branches, bent by a strong northeaster. The big coniferous trees need all the root power they can muster in unfrozen, sandy soil to withstand this onslaught of nature's destructive power. The phone became my weapon to plead and threaten the sewer contractor. He promised that Christmas Eve morning he would do the backfilling–and he knew he was going to have another phone call. He kept his promise, and none too soon, as the anticipated gale arrived on Christmas morning. In retrospect, I don't know whether filling the trenches with the loose soil would have helped at all, but it's good the excavator did not make this observation after having been so dilatory in finishing the job he had contracted for. As a general observation, it is maddening to drive around the county and observe the many instances of test hole mounds in woods and fields which have never been backfilled into the trenches.

Thanksgiving weekend I spent almost entirely in the office translating my envelope noodlings into a preliminary architectural floor plan and creating the exterior elevations. I precisely located the building tucked into the back side of the natural clearing, which had so influenced my initial enchantment with the site. The topography and tree locations depicted on the survey dictated the placement of the footprint and the size of the building. It just fit, squeezed between flanking cedars on the west side and hard upon the required side yard setback on the east.

I had studied the ordinances to determine their definition of high-water mark to factor in the 75-foot setback. No problem, I still had at least 20 feet from the rear of the house to the tree line. No problem that is, until I submitted the drawings for the County Zoning Permit. I felt it was important to be there when the zoning administrator did his site inspection. I would endeavor to do this for any client, to nip in the bud any off-the-wall interpretations inimical to his interests.

It was with one of the assistant administrators who I met on-site early in December. All business, this young woman was, a year or so out of college. She immediately took issue with my interpretation of the high-water mark, being at that point the waves had eroded an embankment into the grassy dune. She pushed the high-water mark back by 20 feet where there was an irregular line of dogwood bushes and one small cedar tree. My entreaties that this would push the house back into the tree line, and that I did not want to cut down any of the beautiful cedars and pines to build the house, were voiced to deaf ears. Not only did it not arouse any sympathy, but elicited the comment that it was her opinion when developing property one should never take into consideration trees because they could always be replaced.

It was fruitless replying that I was buying this land because of the gorgeous mature trees. She looked at me stonily and repeated that this was always a mistake. This was a philosophical debate going nowhere, and I have learned it doesn't help your cause to antagonize a building official from whom you are desiring a permit approval. She stated that her boss would agree with her on the position of the high-water mark. The Wisconsin Department of Natural Resources (DNR) gives them their guidelines so that avenue of appeal would have dead-ended. Anyway, I wanted to build a house, not enter into a quixotic debate with officialdom, so best to bite my tongue and bow to the inevitability of the bureaucratic will. A contest of wills should only be entered into when victory is attainable. When departing, she recollected it was the sixth time that year she had been out to this site with people interested in it.

Thank the Lord for mosquitos, clingy bushes, swamps and stormy weather. This property was a "sleeper" with its attributes hidden to the layman's eye. I just happened to have the perception peculiar to our profession to visualize the potential of its transformation into a homesite of exquisite natural beauty. The professional acuity of site potential I possessed was enhanced by my appreciation of the natural world in all its diversity. The meandering creek and its floodplain, the slough formed by the oxbow cutoff, the significant topographic changes, the swamp; each may have turned off prospective developers. To me they were characteristics inherent in the mosaic of a natural microworld supporting a great variety of plant and animal species to be discovered and enjoyed in every season. All of these features are profoundly influenced by the true seasonal microclimate found along the shoreline of Lake Michigan, where spring comes two weeks later than it does a quarter of a mile inland; where fall stays warmer later; and where winter's wet snow inland can be the shoreline's rain; or with a peculiar onshore wind, can contrarily create an unusually rapid accumulation of snow.

The administrator's rejection of the high-water mark location necessitated a rethink and redraw of the site plan. After she departed,

I took some accurate measurements from the high-water mark to the rear of the house. Fortunately, the rear wall would still be about five feet from the nearest cedars. With their lower limbs trimmed up, the planned walkway could become a pleasant natural arbor to the rear door and past the dining area box bay. I resubmitted by mail and had a zoning permit returned within a week.

# Chapter 4
# Design Dilemmas

December began a serious application of my nights and weekends at the office beginning to prepare the detailed construction drawings of the house. After several weeks of not getting much accomplished, I admitted to myself there was always something to do at the office more demanding of my attention. Regardless of when it was, my project seemed to be playing a sorry second fiddle to the never-ending pace of work to be done to earn a living. I decided the only way I was going to get these drawings done was to do it at home. On Christmas Day I dismantled one of the drafting tables, hauled it home and reassembled it in the second bedroom. Now I could devote my truly leisure hours, whenever they were, to drafting a set of plans. Two months and seventeen sheets later, the job was complete with the help of our senior staff architect. I had crafted the images of my dream onto paper ready for contractor pricing.

The central portion of the structure was framed entirely of unfinished western red cedar, heavy timber trusses held in place with bolted steel plates. The heavy timbers not only supported the roof, but also created a beautiful geometric tracery of functional form visible throughout the interior… and from the outside through the double height bay facing the lake. The cinnamon-colored cedar timbers would exude an aroma within that even to this day is sensed when entering the house. The rear one half (kitchen and dining area) of the central house is under a loft deck resting upon the bottom chord of the trusses. The loft looks down upon the living area with its stone fireplace and out the large bay onto the beach and the unbroken blue-water horizon. The massive great room fireplace structure and chimney column also surrounded a second fireplace and flue, being the one in the master bedroom wing. This bedroom was to have its own bay and window seat fronting on the beach. The opposing offset bedroom wing contained not only the guest bedroom and its bath, but also the utility room.

The constraints of the trees limited the length of this wing. I was faced with the choice of providing stair access off the entry hall to the loft over the central house and the attic over the west wing, or providing a fireplace in the guest bedroom. There simply was not space enough for both. Bowing to the practicality of living in the house, the third fireplace was omitted from the design much to my disappointment. It's not that a fireplace was anything but a luxury for a seldom used room. It was simply an aesthetic element which would have removed the guest bedroom from the ordinary to the unusual and special. Outside, the third fireplace meant sacrificing two large cedar trees close to the end of the wing. This fireplace ended up being about the only sacrificial compromise in the design.

The view from the bed in the guest room was to prove as spectacular as I envisioned when drawing. The full-height bay framed a centerfold vista of the tree line along the shore out toward the point and the lake beyond.

Although the house fronts on Lake Michigan's west shore, it actually faces south as it is tucked under the point of land around which the littoral falls away toward the north again. An enviable circumstance of the south-facing house is its large expanse of window glass on this side. Even on the coldest day in January, the weak refracted sun in its low transit across the southern sky suffuses the interior of all three principal rooms with natural light and solar warmth. By 9:30 a.m. in the morning the heat goes off and remains so, until at least 2:30 p.m. Would that the sun show its smile regularly in the winter season of abbreviated daylight and unrelieved cold temperatures.

Digging in on the drawings and creating the details provided a rush of enthusiasm within me. I couldn't get at the drafting table fast enough or long enough. Every spare moment, evenings and weekends, was devoted to the plans. I wanted the architectural features and details of this small building to be perfection, at least in my eyes initially, and hopefully when completed to be appreciated by others.

Working within the creative end of the architectural world, it has struck me again and again that you often sense when you have given birth to a gem. It's a sixth sense that anyone in any creative endeavor has, whether it be architecture, painting, dance or whatever, which rings an intuitive bell within the creative psyche to say, "I think you've got a winner." Of course the proof is not on the drawing board, but in its critical reception when completed. Yet the creator's educated intuition early on is what gives impetus to inspiration and enthusiasm in completing the design work to the point where it can be built by the craftsman, danced by the performers or shown in a gallery. It was not a process evolving in a vacuum. Opinions were solicited at home to become aware of Jan's needs and perceptions because she, indeed, was one-half the client to satisfy. Informal discussions were held in the office to get others' professional opinions—a confirmatory sense of the validity of various aspects of the design I was noodling on. Late in January I had enough done that I could interview contractors using a rather detailed set of documents to give them an insight into the complexity of the project, and the idiosyncracies of the owner-architect for whom they would be working. I wanted to select a firm not only by price, but because we had the ability to communicate, and the contractor to be cooperative. The interviews were to make clear the building was to be built my way with no deviation from the plans. I offered that I was open to suggestions beforehand but did not like surprises after the fact. I could not be there each day to monitor their work so I wanted to find out if they were willing to play by these rules; and if they could start in May and proceed until finished, myself desiring occupancy in the fall.

I drove up to Sturgeon Bay and interviewed four contractors who had been recommended as being capable of doing the quality work I expected. One contractor bristled at the suggestion that I would be telling him how to do it. I suggested that it was the drawings that would be telling him how to do it. He said he would follow the plans unless he felt he could do it better his way. Wrong! That's about all it took to conclude that this was not going to work amicably. Another said he would be conscientious in following the drawings, but, due to other work, could not guarantee carpenters on the job until sometime in July. Although I liked this man, his inability to service the job with carpenters until midsummer would mean a delay in occupancy. The third showed up at his office an hour late for our meeting and had much the same attitude as did the first. Somewhat disheartened, I drove thirty-five miles north to interview the fourth. All the negatives, that had kept me from becoming serious about this sort of busman's holiday in the first place, kept welling up in a nightmare-like vision of the possibility of this becoming a knot-in-the-stomach aggravation.

Don, "Everyone calls me Yukon," listened to my rules of the game, attentively reviewed the drawings, and stated though the house was not large, it was very detailed and unique to the area, and he would very much like to do the job. He said they would faithfully follow the plans, would begin work at the end of April, and work every day until it was completed. He also said his partner John would probably be the lead carpenter on the job. This contractor was large enough to have the benefit of more than one carpentry crew.

We chatted about subcontractors and, as it turned out, Don's firm did just about everything except the mechanical and electrical trades. He was the one who suggested we get Roger (the Roger of the answering machine) to bid the mechanical subtrades, as Roger was located quite close to the house.

Don did the excavating, grading and concrete work himself. John's son was a very skilled mason, both brick and stone. John and the carpenters not only did the rough and finish carpentry, but also the insulating and drywalling, the roofing of cedar hand-split shakes, the siding of cedar shingles and the wood flooring.

Performing all these various trades with their own men was very attractive to me, as it meant much better control of scheduling and performance than the more typical live-or-die reliance that general contractors have upon subcontractors. This unfortunate dependence makes general contractors' schedules unreliable and their business lives frustrating.

I parted the meeting with an understanding that Don's firm was going to be the contractor. When the plans were complete, I would send them up for his labor and materials takeoffs, with the agreement that he would get multiple bids on those areas of work to be performed by subcontractors. He also agreed to get more than one bid on all the building materials. I was to be privy to all detailed labor and materials takeoffs.

The kitchen and bath cabinetry was not to be part of Don's work. Looking for millwork houses locally was not easy. I interviewed one in DePere (a long way away) which Don suggested, but they wanted to build the cabinets their way, not mine, so one of their standard cabinet designs could be used. They reluctantly agreed to do it my way and priced it both ways, either of which was out-of-sight.

I had at hand what I considered to be the finest cabinetmakers in

_____ Design Dilemmas

the world right at Palmer Johnson, but PJ is accustomed to a billing rate on yachts at least double what millwork houses servicing the building industry would charge. Anyway, they really didn't want to be making kitchen cabinetry. Their ship's carpenters were always busy doing what they do best. One of PJ's management was listening to my tale of woe concerning my inability to find suitable local custom millwork. He said a former employee had a small shop in town and was very skilled.

Drawings in hand, I found the shop behind his house and poked my head inside to a world of airborne wood dust and the screech of tools. I was greeted with a friendly smile half hidden by a very powdery full beard. All over the small shop were cabinets in various stages of completion and stacked in the corners were guitars in process. Later I learned Dan makes guitars as a hobby. Talk about a busman's holiday. On the other hand, it was certainly indicative that the man loved what he was doing–creating beautiful things out of wood.

Dan said he would build the cabinetry to the appearance I intended. The rough-sawn cedar exteriors of the cabinetry were to match exactly the central house's interior wall facing of vertical six-inch lapped, square-edged, rough-sawn cedar painted white to resemble whitewash. All fabrication of the cabinet exteriors (and the wall paneling, for that matter) had to be done after the boards were painted on three sides to prevent inevitable shrinkage from exposing the raw wood on the edges.

As it turned out Dan's price was one-third that of the DePere firm, fabricated, finished and installed. But even better than the price was that I knew instinctively this craftsman was going to give me exactly what I expected, at least in quality. Timing was a concern not readily confirmed because everyone always says they can meet the schedule. The long lead time was not very beneficial because he had to wait for the house to be framed before he could field measure. Except where a cabinet has a finished end, accurate measurements are crucial to a successful installation.

The success of my contractor search drove me to greater dedication finishing the plans. Yet it was nearly February and we still had not closed the real estate deal. Not being able to reach George, I began calling others of his surname in the town where he lived until I turned up his son who gave me a phone number in Texas. George was in no hurry. He said he had intended to call when he got back home. We set a date in early February to meet at an attorney's office in Sturgeon Bay to close.

By 9 a.m. that day the closing was over and the deed recorded. Jan and I damn near flew out to the property feeling the headiest sensation imaginable with the success of a mission accomplished, against what had at times seemed to be enormous obstacles and frustrations. The property lay under a deep covering of snow. Fortunately, someone had been along the track on a snowmobile so we could walk on the packed snow track. Muggs, never having experienced deep snow, barreled off the packed track and disappeared into two feet of soft snow. Struggling to get back to something solid, the only thing that kept her afloat, so to speak, was her broad carriage—she's built like a barrel. The short legs of the Scottie couldn't reach terra firma. So I picked up Muggs and set her back on the trail to shake off the snow and her disconcerting experience to barrel onward somewhat chastened and wiser. The exhilaration of walking along winter's path through the still, snow-mantled forest on this property—our property—was cosmic. If I could have, I would have told the trees that they now had a steward who loved each of them and was going to watch over them, and nurture them. But at that singular moment in time I had no inkling what my love would become, and the care I would in the time ahead come to lavish upon my trees, nor the anguish I would feel at their loss.

At the end of February the drawings were complete and sent to Don for pricing. Three weeks later we borrowed our friend's home to go over the bids with Don and John. It was no surprise the prices were more than anticipated... a lot more! Architects are their own worst enemies when it comes to this sort of thing. Whereas the layman thinks that the architect would have an excellent handle on cost control on his own project, if no other, it seems it just doesn't work out that way. The architect is conscious of all the enhancements; nifty items and features which elevate the home from the ordinary to something special—and includes all these premium details. Great for the aesthetics, but devastating to the pocketbook. It may have been that this architect was guiltier than most, succumbing to this architectural affliction. It was not that I had been ostentatious or pretentious in the design—only that premium materials and good things cost! The premium list was long and the meeting longer, weighing the pros and cons of substituting.

A roof of hand-split cedar shakes, the exterior walls of cedar shingles framed within one and one-quarter-inch-thick painted wood trim (corner boards, window and door surrounds, eaves and gables, the five bay windows and more), all were very labor intensive.

Exterior walls were framed with 2 x 6s rather than the more typical 2 x 4s because I wanted the extra depth of insulation and a more substantial appearance in structure. Fireplaces, hearths, and chimney columns, both interior and exterior, called for weatheredge limestone. Oak plank flooring was desired throughout the central house. The wall surfaces throughout the interior were to be cedar boards. One and three-quarter-inch-thick paneled doors were specified with the best grade of polished brass hardware.

The heavy-timber roof framing of old growth, western red cedar was necessary to get the uniform cinnamon color. All the timbers were to be individually wrapped to avoid watermarking during shipment and construction. A house with twenty-nine corners… and on, and on. There were some savings to be attained in items that did not adversely affect the architecture, of which we immediately took advantage, albeit minimal in benefit.

Here I was, the spider caught in his own web, trapped in the typical situation of my clients—the bids forcing their dream home to be compromised by the harsh reality of having to pay for it. What made things even worse was Don's pricing did not include the windows, the hardwood floor material, finish hardware, light fixtures and stone, all of which I was purchasing through my own sources, calling in some IOUs from over the years. Nor was the cabinetry to be done by Dan included.

Yukon indicated he could plaster the interior walls less expensively than installing the cedar siding. That afternoon we drove to a house he was completing and looked at the decorative plaster finish. We quickly decided the plaster would be a superb looking wall surface of just the right "rusticity" throughout both wings and in the loft, leaving the whitewashed rough-sawn cedar only in the central house—a decision I have never regretted.

The drive home that evening was somewhat somber with the avenues of compromise clearly in focus. But, and it was a big "BUT," the substitutions suggested which offered significant savings also would significantly alter the home's architectural features and ultimate character. This took some real soul-searching. Do we adjust ourselves to a larger budget or compromise my vision of this special design, this gem I felt I had created. Believe me, it is much more difficult philosophically for the creator to compromise than it is a client. Architects are always trying to convince clients not to compromise aesthetics. Despite that I am by nature not a big spender, and in fact did not have a hidden money tree at my disposal, I decided before reaching home to proceed essentially as shown on the drawings—without any further reduction in quality. This was going to happen once in my lifetime, and I intended to give it my best shot. If I settled for something less, I would forever be living with and looking at something which I wished was something else. I called Don upon reaching home to let him know of my decision. His retort was, "I knew you were going to do it."

Jan only commented that this was turning out to be the world's most expensive doghouse.

Yukon insisted upon calling Muggs "Rover," so Jan always addressed Don, "Klondike." He never could quite figure out how to handle this bit of chiding, but Rover it remained. As Muggs couldn't defend her name, Jan felt it only fair to be the needle in her behalf. And so it stands.

Don had given me all of their materials takeoff sheets. After having spent several hours reviewing these quantities, I suggested to Don that I meet with the person who did the takeoffs as it seemed there were duplications and overages. He graciously said that would be fine, but qualified that it really wasn't necessary as I would only be paying for what was purchased and used. As our deal was one based upon time and material, not to exceed a certain agreed upon limit, it made sense to me to reduce the limit to an authentic representation of what was necessary. It has been my experience that for some reason cost always seems to rise ultimately to whatever is the limit. My time spent was well rewarded as the materials cost went down by $8000.

The finishes of the interior, simple and somewhat rustic, had to be complemented with furniture and fabrics selected and coordinated with the overall plan in mind. Jan had several antiques which made wonderful and useful features within the central house: a pine dry sink, a grandfather clock, a spinning wheel and a butter churn. We selected and purchased a large oriental rug from which the colorway flowed throughout the central house. Navy window seat upholstery and russets on bench seating were selected. The cushions on the white Bar Harbor wicker chairs were to be a pattern of both colors.

The dining table nearly stumped us. We looked the world over (it seemed) for a 3 by 7-foot pine table—new or antique. If the size wasn't a problem, and it was, the construction was shoddy on those we found.

The savior of this dilemma became Jim, the superintendent at Palmer Johnson, who had retired a year or so before the house was begun. Jim, a ship's carpenter by trade and a cabinetmaker by choice in his retirement, said he wasn't interested in doing the cabinetry for the house but would enjoy making the dining table if I would provide him with drawings and details. Jim is busier in his retirement than he even was at the shipyard. His well-tooled shop in his barn is heated by a wood stove, and it is seldom he can't be found there crafting something or other. This heavy pine plank dining table with its turned legs challenged his skill, and it turned out to be a masterpiece. It isn't varnished. I had one of the painters at PJ stain it, beat it with a chain and apply four coats of wax over the raw wood. When Jim saw the finished table his heart plummeted. He asked me how I could have ruined this beautiful wood. My explanation of antiquing seemed hollow and lame at best. To this day he hasn't let me forget what I did to his table.

We had selected a narrow, dark oak sofa table with inlaid beveled glass under which slid two upholstered benches. The manufacturer did not make a coffee table to match, so I called Jim and said I had another project for him. When I described the table, he asked if I was going to ruin it like I did the dining table. Assured that I wasn't, he said I should get the drawings to him, and he would get going on it. I gave one of the benches to Ted, the painter, to match the oak finish. The result was another masterpiece of craftsmanship in wood, not rustic, yet traditional in design. I only wish the sofa table, which came from one of the finest furniture manufacturers, was made half as well. Jim's coffee table could withstand the scrutiny of any discerning buyer in any showroom.

During these months while the house was being designed and was under construction, Jan scoured the newspapers for sales. She haunted the showrooms of department stores and specialty shops, purchasing the myriad of things required to furnish and maintain the home. Our basement became a warehouse, filling up with wicker chairs, rugs, cartons of utensils and dishes, cushions, bedspreads and other necessities. The furniture store where we purchased the couch, lounge chair, beds and mattresses during a great sale offered to store these items until we needed them at move-in time. This was fortuitous as there was no way the basement could hold these along with the rest of the stuff sharing space with *Aurora's* off-season storage needs. Besides, we would never have made the twists and turns down the stairs with this bulky furniture.

The last official approval needed was the building permit from the Town of Sevastopol (a township in some states). I made an appointment with the town building official, who was an appraiser in Sturgeon Bay, asking if he could review the plans while I waited so I could answer any questions.

"Well, let's see the drawings."

I don't know whether it was shock or dismay which began to register on his countenance as he leafed through the seventeen sheets. He said that he'd never seen a set of drawings like this before. With several questions about electrical and one concerning plumbing, he rolled up the drawings and candidly stated he was sure I was not trying to get away with anything. We chatted about sailing, which he was evidently interested in from the photos on the wall, while he wrote the permit. Fifteen dollars and fifteen minutes later I had my building permit. He said he would be out at the house when the electrical and plumbing were roughed-in to have a look. I never heard from John that any inspections were ever conducted by any building official during the construction. Some things are pretty loose out in the rural areas.

# Chapter 5
# Getting Underway

With the snow cover rapidly disappearing, Easter week saw the well digger at the property, having had his lecture about being careful of my trees. This was somewhat worrisome as the large rig and its support vehicles had to back up the narrow winding track, and the rig had to be erected through the tree limbs to drill. Because of groundwater contamination, the well had to be cased down 140 feet. This, in effect, was the minimum dig. Ours ended up at 440 feet.

Springtime awakened the holding tank installer who followed on the heels of the well digger. To enable pumping out when the system was in operation the only possible location for the 2000-gallon steel tank was next to the driveway, shoehorned in-between several trees. He received another admonition regarding trees. Now we were starting to get into the big-time tree whackers—excavators and their machinery. There was the probability of needing to blast into bedrock to get enough depth for the tank with its required concrete and earth cover. Fortunately for this worrier, blasting did not become necessary with only inches to spare.

Prior to Don and John beginning work I wanted to get the electric service in to the building site, if for no other reason than they would be needing this for their power tools. The utility said they would have to cut a ten-foot-wide swath through my forest to bring in their overhead line, and to make matters even worse, would have to install a pole because of the distance to the house. This was absurd, a ten-foot-wide path for a one-half-inch wire route.

It seemed like there was a conspiracy against my beloved trees. Overhead service was obviously out of the question. "How about running underground?"

"Oh yeah, it can be done, but only for 100 feet."

"Why only 100 feet?"

"That's company policy."

"What if I agree to pay for the added underground run?"

The sales rep was getting a little testy after I told him what I thought of the requirement for a ten-foot-wide clearing. "I told you it can't be done because we're only allowed to go 100 feet."

"Even if I pay for it?" I incredulously asked again.

He grumbled that he'd never heard of it being done before.

"How can you find out?"

"I'll have to call our main office."

I suggested we mark the desired service route so he could determine exact distance and equipment feasibility on the chance the underground service would be approved. The wire course snaked through the trees and up the side of the driveway, beside the holding tank and diagonally to the house along the intended route of the sewer line. The trenching equipment only required a 30-inch width to operate, so it wasn't too difficult to avoid tree roots.

By the time I got done with the utility rep he was more than convinced about my earnest concern over the trees. I had hoped to make him aware of the wrath he and his company would suffer through if the trees were damaged. It is my experience that public utilities are notably insensitive toward the environment when it conflicts with line placement and maintenance. He evidently thought he was dealing with a certifiable nutcase as he commented that if he had anything to say about it (which he didn't), he would never install my service as he did not want to be responsible for damage to trees. A real confidence builder!

A week later he called and told me they had permission to install the underground service if I would sign the contract and give them full payment up-front. Whatever the cost, obviously I had no choice, as is ever the case with any utility monopoly. Of course I could have settled for the free overhead service, but by this time, it must be fairly clear that this option never was for this very particular home builder. They agreed to do the work within a week after receiving the contract and payment.

Amongst other things, I never gave the electric service further thought until I had a call from my reluctant utility company rep. Some intuition told me I didn't want to hear what he had to say. Sure enough, a very shaky voice on the other end of the line said their crew had hooked a large root and toppled one half of a good-sized double-trunk hemlock. Knowing the route we had staked, I did not see how this could be. Slowly the facts surfaced. For some reason, the sewer contractor had volunteered to excavate the trench for the power service crew as they were approaching the holding tank. This was crazy. A backhoe with a 24-inch-wide scoop, digging the trench for the wire. What made it all the more inexplicable was that they had diverted from the staked route by over 15 feet. The rep was most apologetic and said his company would give me credit for the damaged tree. What can one say? The badly scarred remaining hemlock trunk stands today as mute testament to this bungled work.

By mid-April the surveyor had precisely staked the footprint of the house, and Don had (as we had agreed) protected all the trees along the sides of the driveway up to the house and in the vicinity of the excavation. I had specified that the trunks of all these trees were to be surrounded with 4 x 4s. When pricing the job, Yukon commented that eliminating this tree protection endeavor could be one way to save some money. It was then he got THE lecture, by now a recorded message off my tongue. I also added that I was going to insist he personally do all the excavating and grading, or we didn't have a deal. This was not onerous to Yukon as he enjoyed maneuvering his bulldozer as I do sailing *Aurora*. He prided himself on his skill and deftness handling the tracked behemoth. With my harangue about the destructiveness typical of excavators and the efforts I had him take to protect the trees, there was no way he was going to disappoint me. His pride wouldn't let him, and he didn't.

On the 15th of the month, Yukon and I had to find a route to the beachfront for construction traffic as the house was being built on the pathway of the old trail where it crossed the small meadow. The only feasible route required cutting a large (12-inch) horizontal branch of the magnificent red oak located southwest of the house.

"OK, so it's only one branch."

Anyway, there was no alternative as I had precluded staging of materials anywhere but on the grassy beachfront. We also staked out the barrier to be erected around the oak tree at its drip line to prevent construction traffic or storage within its circumference. He wanted to use the new type plastic snow fencing which is absolutely worthless for long-term durability. I insisted he use a wooden snow fence. Ringing the oak tree with this enclosure took away a great deal of needed staging space as the lower branches extended nearly 30 feet from the trunk and were, and I might add "are," only five to six feet above the ground. Sure as hell someone delivering materials down to the beachfront would have hooked one of these low branches trying to get in or out. Even Don admitted that it was the only way to keep the tree safe from harm, though it created somewhat of a maneuvering problem. These protective measures were 100 percent effective. Somewhat abused and worse for wear, the 4 x 4s around the tree trunks, and the snow fence circling the oak tree, were the last evidence of construction to be removed at the end of the job.

Once Don became aware of "the program" he joined in with a will and even began making his own suggestions. He must have transmitted to John and their employees a like degree of care that was expected of the folks working on the job. They even became conscientious about using the dumpster to deposit their soda pop cans and lunch wrappings, rather than strewing them about the site. I never did learn what they did for a toilet. A pit in the woods would have been worse than torture with the voracious mosquitos needing only a second or two to find and ravage a dormant suspended bottom.

Though precise by nature, crossing my t's and dotting my i's (and then checking to see that I haven't missed one... I'm a worrier, too), I am a person of boundless enthusiasm toward whatever endeavor I have committed my energies. I'm not into eight different things at the same time, so retaining focus is normally effortless. The moment of realization was at hand, toward which all creative, technical and logistical considerations had been planned. Everything was in place for the symbolic shovel of earth to be lifted.

It was with some feeling of ironic sadness that I stood out on the beach and gazed at the undisturbed natural scene facing me. In a short period of time I was going to permanently alter this beautiful vignette of forest and meadow which had evolved over the generations. I know that the reader is certainly wondering what the emotion is all about, "For God's sake, you've done everything possible to preserve the environment, so get on with it."

I simply want to touch upon this eternal conflict that goes on within my psyche. My desire to leave undisturbed the incomparable and irreplaceable beauty of the natural world tugs against the converse necessity of my being part of the building development process to earn a living. Practically speaking, I always rationalize that if I don't, someone else will; and while doing it, I philosophically try to do the best I can to create something worthwhile in harmony with its natural surroundings.

This omnipresent dichotomy of emotions did nothing to dampen the palpable exultation I experienced upon reaching this milestone on my building journey–the one I swore I would never get involved in. I felt great about the design and confident I had found just the right relationship with the craftsmen who were going to put it all together.

# Chapter 6
# Our Participation

Ten days later Jan and I drove up to inspect the progress. Ten days was to become about as long as I could endure without seeing what was going on. The footings were in, and the masonry foundation walls were about three-quarters complete. We observed that the perimeter of the house looked so small. It has always mystified me why buildings at this stage of construction appear to be much smaller than they truly end up to be. Have faith in the plans. Shortly the foundation walls were to be finished, and the mounds of excavated soil (sand in this case) could be backfilled around the perimeter of the walls. The site would then be graded out and excess material hauled away. By the next visit, I expected the floor framing and rough floor would be installed. Materials were beginning to fill up the site: framing lumber, structural steel, plywood sheathing, concrete block and the inevitable dumpster.

On the trip home, we detoured through Fond du Lac to visit the stone quarry from which I wanted to purchase the weatheredge limestone for the fireplaces and chimney. This quarry had previously supplied stone to jobs of ours, and I was confident of their quality control. We drove into the quarry to select the unquarried stone for the two hearths. Each hearth was to be one piece, the larger being 8 feet by 4 feet by 4 inches thick with at least two sides and top being the natural, exposed rock face. This took quite awhile as natural rock faces are not straight and uniform, nor do they normally have 90-degree corners. The quarry owner said they would ship within two weeks. Although we would not be ready for it by any stretch, I would rather have it on-site than Don complaining that my stone supplier was holding up the progress.

This early delivery was terribly fortunate as the next time up, by chance a day or two after the stone was delivered, we realized immediately the color was wrong, too gray (not being the tans we had selected in the quarry). Also, the hearthstones were each in three pieces. So much for quality control. I burned up the phone lines to Fond du Lac.

The owner of the quarry agreed to reship and when confronted with the impracticalities of loading the stone to be returned, he said we should keep it. Don got a lot of free weatheredge limestone as a consequence. I wonder what the quarries reaction would have been to my refusal of the stone if I had not been an architect with whom it had done prior business with the expectation of more. At any rate, this illustrated the axiom that one should inspect a product prior to paying for it.

During the periods between visits, we were very busy at home with related activities. Living in a huge metropolitan area, such as Chicago, has it advantages and disadvantages peculiar to building something. Products are readily available or attainable, but they usually cost more. When it came to getting specialty items, the contractors and suppliers in northeastern Wisconsin were often at a loss. For instance, they could not find a supplier of square-head bolts. Hex-head bolts weren't acceptable to my finicky sense of an authentic old look. *The Illinois Manufacturers Directory*, which I use in marketing architectural services was invaluable. After several calls to bolt manufacturers in the Chicago area, I discovered that square-head bolts weren't ordinarily manufactured any longer, but I did find a manufacturer with a stock of the old square-heads. Probably thinking they were fortunate to find a taker, they not only took the order for 576 three-quarter-inch nuts and bolts in various specified lengths, they even delivered them to our office. This was great, but the cartons were so heavy two of us could barely lift them. I had asked John if the rough-sawn timbers were going to come full-cut. Inquiring of the lumberyard, the reply was "full"; that is a 6 x 6 is truly 6 inches by 6 inches, not typical of finished lumber where a 6 x 6 is actually 5 1/2 inches by 5 1/2 inches. All my bolt lengths were sized to the full-timber dimensions.

So many times when something gets fouled up, it just gets to be a God-awful mess trying to resolve the problem with one screw-up following another. Ordering the heavy timber became a case in point. The next time up, John asked me to visit the lumberyard to inspect the timbers prior to their being delivered as he didn't think I was going to be satisfied. Despite my having been very specific about color and wrapping, the timbers shipped were not the uniform cinnamon color, nor had they been wrapped, so many had become watermarked. The supplier had agreed to the conditions in his pricing and had not followed through with the mill where they were cut. They expected at

 least a ten-day delay getting replacement timbers, which I thought to be terribly optimistic. It turned out to be three weeks! But this didn't matter because Door County was suffering through two solid weeks of rain that precluded any job progress after the floor was installed. The second time the mill got the color and wrapping right, but all the timbers, though rough-sawn, were sized to the smaller dimension of smooth-sawn lumber. This would not have been a problem, except the bolts were all sized for the larger full-timber dimensions. The result was all those bolts associated with connections exposed to view (the majority) were too long to look good in place. It was clear we had a loser going to again demand replacement timbers, this time full-cut. I didn't feel the odds were in our favor to reorder again and have any hope for a timely delivery. Besides, the other aspects of the timbers were satisfactory. So, I called the bolt manufacturer and asked if I could reorder approximately 400 of the bolts. "You bet." The originals were on the job, and John had offered to pay for the replacements... I'm sure with a backcharge to the lumberyard. Now this was a solution which short-circuited the God-awful mess.

I was getting pricing on all the ironwork associated with the wood roof trusses. There were fourteen variations of plate connections, all of which had to be matched and drilled for bolts in facing pairs; and in the case of the ridge beam, connections in quads. In some cases, the fabrication of the connections also required welding. All were to be painted flat black and delivered to the site wired together in numbered pairs (or quads). The price of the ironwork in the Chicago area was astronomical, so I tried a small steel fabricator in Sturgeon Bay with dramatic success in price and performance.

Copper gutters, downspouts and fittings were ordered from a fabricator in Chicago and shipped to Don at his shop.

I spent a couple of evenings at the workbench in the basement sampling white paints on the interior rough-sawn cedar siding, subsequent to the loft deck being installed with its bottom side stained opaque white. Though the color was acceptable, it had an objectionable sheen like paint or plastic. This certainly wouldn't do for my whitewash look on the paneled walls. I was searching for a finish that would do the job with one coat, a tall order because of the absorption of any coating into the rough-sawn boards. Otherwise, this prepainting deal was going to end up costing a fortune. After several experiments I got lucky. One coat of flat enamel provided exactly what I was looking for, with remarkable color retention on the surfaces of the wood.

I felt it prudent to explain to John that the carpenters must not get the prepainted panel boards dirty or finger-marked during installation. The only way a rough-sawn surface would be cleanable would be by applying another coat of paint after installation. The second coat of paint would completely alter the appearance, thus destroying the look I had tried so diligently and successfully to achieve. I suggested they wear gloves to do this work. Incredulous, his mouth opened to say something humorous, I presume. Nothing came out but a withering, "OK."

Jan and I visited the Merchandise Mart in Chicago to select the decorative light fixtures. Raw brass finish, showing all the solder joints, became the design theme for all the fixturing, both exterior and interior. The same tiring day we selected the upholstery fabrics, pouring through hundreds of samples in the fabric showrooms until persistence paid off, and we had exactly the colors and textures we were seeking. The incredible volume of fabrics from which to choose amazes me, and all the more so, because you may just be lucky enough to find one thing that sings, after searching for hours through the racks.

Since early on I have always had a fascination with barrels; old wooden barrels with iron hoops. When I was a boy, I would scavenge construction sites for empty nail kegs which I brought home and stored under the porch, only to forget about them. But now I wanted a rain barrel. Calls to several cooperage houses in Chicago resulted in nothing. Barrels in the cooperage industry today are fiber drums or steel drums. One last call netted a gold mine. This supplier had twenty-seven oak whiskey barrels left and wanted $15 a barrel.

# The Paradox of Paradise

Realizing that I may have been searching for an endangered species and had found a bargain, I left within the hour to pick them up. It's good I was driving a big automobile because these were big barrels with rusty iron hoops, each weighing 100 pounds empty. Selecting three undamaged ones, they were rolled out to the car. One barely fit in the back seat and as I cringed, the second was jammed in the front, except it didn't leave any room for me. The third was in the trunk with the lid tied down. I drove back to the suburbs squashed sideways in the driver's seat. Despite the barrels having both ends intact, and their bungs solidly in place, the inside of my car smelled like a distillery. A month later the aroma was still hanging around. Obviously, I wasn't going to be able to transport these to Door County in the car. I heard about the damage the one in the front seat did to the dashboard, and what its rusty hoops did to the leather, for a long while. I stored them in the garage until I could borrow a friend's van—he never objected to the aroma of sour mash. I had forewarned him. The barrels went to Jim who removed an end on two of them. The one to become the rain barrel had holes drilled in the bottom and a wire basket was made to fit down into the barrel to catch the pine needles and cedar leaves, as the collected rain water drains out the bottom. Jim made an oak lift-off cover for the second barrel to be used as a garbage receptacle. I found a plastic garbage can that fit snugly inside as a liner. To this day, each time I lift the lid off, my nose is tickled by the sensation of whiskey.

The house does not have gutters except at the back

entry and across the open face of the woodshed. The copper downspout servicing the woodshed's gutter has an elbow at the bottom into the rain barrel. The cold climate necessitated that the barrel be upended in the winter, when the elbow is attached to a downspout extension to flush out at ground level for six months of the year.

When the lower hoops started dropping off the garbage barrel after a year or two, it occurred to me that the wood staves had shrunk and I'd better do something, or I was likely to have a pile of barrel staves on the ground instead of a barrel. Filling the barrel with water did the trick as the water squirted out numerous leaks, until an hour or so later when the staves swelled against the hoops and the barrel became watertight again. It reminded me of launching a wooden vessel after it has been out of the water awhile. It has to stay in the slings or on the ways, with its planking swelling up, or you could find your yacht sitting on the bottom with as much water inside as out.

The monsoon in Door County ended, and the wall framing was started in earnest with five Swedish carpenters on the job. We drove up on a Thursday night so we could be out at the building site early the following morning. Friday turned out to be a picture-perfect day with a sunny sky and a gentle onshore breeze. While I spent the morning with John, Jan and Muggs parked themselves on the beach; Jan in her beach chair with the water lapping at her feet and Muggs under the beach umbrella, keeping her black body out of the sun. Some dogs search for the sun, this one can find the smallest sliver of shade into which she wedges herself.

This was to become the drill throughout the construction period, with an interlude of a picnic lunch on the beach and then time for Jan to take progress photos. The photos were to become a marvelous chronological record of the construction and a remembrance of the people who so skillfully and enthusiastically accomplished the work. During lunch on the beach, the sounds of saws and hammers were music to my ears as the skeletal frame took shape before our eyes. By the end of that day, the carpenters were beginning to frame the roof. I have always regretted that I wasn't present as John and his carpenters fabricated the heavy-timber trusses piece by piece in place. This was accomplished without the use of any crane, auguring the bolt holes and attaching the iron plates to the timbers, all ultimately connected to a single 34-foot-long ridge beam. When I had previously asked John if he could do this without any equipment, I think I hurt his feelings. His reply was, "I built wooden ships for years. It's pretty much the same."

## Chapter 7
# Tornado

Saturday found me sailing *Aurora* out of Waukegan with the crew—another sunny, breezy day. After the sail, while the Cuba Libres were being prepared, I turned on the marine weather to see what they were forecasting for Sunday. At the time of broadcast, severe squalls were passing across northern Wisconsin with several tornadoes having been sighted, one in Oconto County directly across Green Bay from Door County. The cognitive process took note that Door County would be in the storm's path as it drove eastward on a very predictable course. Hard to relate to on this sunny, high-pressure day we had been enjoying. Sunday's newspaper commented on the tornadoes in several counties in Wisconsin, not mentioning Door County.

Monday morning my breakfast was interrupted by a phone call from John, a not unusual occurrence with the job underway. It never occurred to me that this call was to let me know a tornado had torn across the precise area of Door County where we were building and had left in its path a devastated landscape. In fact, he couldn't even get to the job-site because the Drive was blocked with trees and downed power lines. He said if the tornado touched the property I should expect the worst. I suggested I meet Don and him at the job at 6 p.m., by which time the roadway was to be cleared. Work was very distracted that morning, but being an optimist, I couldn't accept that it could be the worst. Though the odds were in my favor, on the way up I did begin to deal with the possibility of really bad news. Approaching the property from the south along the Drive gave me reason for momentarily misplaced optimism. Yes, there were quite a few downed trees, but it seemed to be an acceptable level of damage—until I reached the bend in the road a short distance from the property. The vista I saw before me was unrecognizable from the same view the past Friday. The tall canopy of the two-tier forest had disappeared. The understory had become the forest. Gone were the stately white pines.

John and Don were standing beside Don's pickup parked on the road, which in and of itself told me something. Their grim countenances expressed more than words could ever have, the anguish they felt for me.

I didn't know what to expect, and what I found sickened my spirit. Walking up the driveway to the house was impossible. None of the trees had broken off. All had been uprooted from the sandy soil. Three- and four-trunk birch trees, where each stem was in excess of 18 inches in diameter, carried with them their extensive platter of shallow roots, now sticking up ten or twelve feet like a plate on end. The big trees took with them many of the cedars and damaged others as they toppled. The white pines and hemlocks lay all across the property, side by side. Where all the beautiful foliage of June had been in place so far overhead, it now was a green impenetrable jungle on the surface, soon to turn brown and lifeless. Trees that remained standing teetered at crazy angles. Two large pines had fallen on the house, crushing the framing like matchsticks. When John commiserated that he was sorry about the damage to the house, I could only answer that the house meant nothing—it could always be rebuilt and was covered by insurance. It was the loss of the trees which devastated me. Never in my lifetime would I ever have the pleasure of seeing another forest here such as it was.

The zoning administrator should have been there. With this ironic twist of fate she would have had the last word. "See, I told you that you should never buy property for the trees."

The beautiful red oak beside the house was almost stripped bare of leaves. A tall ash nearby lost all its branches. The pines which survived had their branches grotesquely twisted downwind—to remain this way forever. One pine lost every branch but a little topknot a foot or two square... a tall stick with a little fluff on top.

I think Don and John thought there was a real possibility that I would be so disheartened I would abandon the building program. It did cross my mind fleetingly. But that evening, when I stood in front of the house-to-be and turned away from the havoc which the tornado had wrought, I gazed across the beach and shoreline out over Lake Michigan, serene and placid, reflecting the lingering brilliance of the summer twilight. It somehow helped me focus on this drastic setback with a different perspective. This half of my vision remained intact and unaffected by the violence so starkly visited upon the forest at my back.

I made up my mind that this catastrophe was not going to deter me, or change my course. I was simply going to have to become creative in restoration of the landscape, building upon what was blessedly left. Don volunteered to cleanup all of Tract 17 so the construction work could begin again. He estimated four days to cut up and burn on the beach what they could, and haul away the rest. I asked him to get rid of all the stumps, as they would be too ugly a reminder to live with forever. I also asked that all the white pines and hemlocks be cut into eight-foot lengths and stacked.

It was at this time I remember asking Don if he knew a quality landscaper. He said he had used one to do several condominium projects and had been very satisfied. The landscaper's business was located conveniently right down the road which was good because as it was to evolve, I was going to keep this landscaper busy for a long time.

After Don and John left I walked around, hopping from tree to tree, to make some kind of assessment of what was remaining. It was impossible with darkness on its way and the inability to see any distance through the foliage. Soon I was sweating profusely from climbing up and over tree trunks and through tangles of branches and leaves, all the while half-heartedly warding off the legions of mosquitos. Disheartened, I sat on a tree to rest... and wept. Never will I forget the awesome scene; the loss of all those trees to which I had so dedicated myself. Even today, when no one but me would ever realize anything like this had occurred on this scenic property, it is hard for me to look at photos of BEFORE. I knew I had to get someone in to cleanup the rest of the property, cut up the large trees and get them out without bringing in heavy equipment to further damage what remained. I called Jim. He and his cousins (who own a sawmill) agreed to do the balance of the cleanup. They would skid the pine logs out to the roadside with a winch and cable and then lift them on a truck to the sawmill. There they would be cut into 2 x 6 framing for a garage I might build someday. Jim volunteered to stack the cut lumber for drying in a shed on his property. Having Jim to do this work was a relief, because I knew he would be sensitive to my concerns and aware that if the trees were important to me before, those remaining were to be treated as precious as gold.

I also met Greg, the landscaper, and went over what could be done to save the damaged and tipped trees. Two large pines were able to be winched back to vertical and guyed with cable. Palmer Johnson supplied me with lengths of wire rigging removed from yachts. Another pine, immediately behind the house, I wanted guyed, but Don insisted his men would not feel safe working on the house with the large tree hovering over them. I gave in as an expedient and have been ever sorry, as this method has worked beautifully. He also wanted to cut the pine stripped of its branches except the topknot. He said it looked ridiculous.

"No way."

I was emphatic that any tree with any green left was not to be cut. God, Yukon was a demon with a chain saw. When he has been back at the house in recent times I have pointedly drawn his attention to that tree. It has blossomed forth from its minuscule topknot with expansive, thick and verdant foliage, to have become in eight years a beautiful specimen; evidently having directed all of the earth's nourishment up its large trunk to infuse a burst of energetic restorative growth. What a joy!

"Look at the tree you wanted to cut, Yukon!"

"Yeah, it's just a tree, Rich."

I did not get an idea of the full extent of the damage until the following visit a week later when the cleanup on our property was for the most part complete. Jim and his cousins were darting branches into a wood chipper and were loading the last pile of logs onto their truck. The sun shown brightly, unimpeded to the forest floor which was not accustomed to such warmth. The back side of the dune, upon which the house stands, was nearly devoid of trees. It now served as a parking lot and a more convenient storage area than out on the beach. Almost beyond belief, immediately around the perimeter of the house, all the trees remained with the exception of the two large white pines that had fallen on the building. The cedars and hemlocks still embraced the home with their branches and provided a wonderful cover of shade. How many times I have expressed my gratitude at this good fortune. The natural environment closely surrounding the building remained virtually intact.

To get an idea what else happened, we drove north from the property along the Drive. It looked like photos of battlefields I've seen. Where there had been solid forest, there was nothing remaining but a messy plain, with jagged trunks standing where the trees broke below the bottom of the branches. Others were stripped with only spikes sticking out where branches had been.

A neighboring house several hundred feet northeast had its interior soaked by the torrential rain, which accompanied and followed the windstorm. Three large pines had punched through the roof. The shoulders of the road were packed with trees and branches with foliage turned brown, awaiting removal after having been pushed off the pavement in the haste of reopening the roadway to traffic. There were hundreds of the upended root wheels with earth clinging to them; the exposed undersides, an interesting, if ghastly, view of the root systems of these large birch and maples. The pines seemed not to carry the roots with them as they toppled. Roots as big around as my calves stayed intact in the soil but snapped off close to their trunks. Other trees had become one-sided as those falling sheared off the branches of their neighbors. The path of the tornado was distinct, its periphery well-defined by the standing forest. Our property was at its right-hand edge. There was little or no damage on the other side of the creek. Having made the comparison between the damage to neighboring properties and ours, Jan commented that we really had a lot to be thankful for, as we still had a half a forest and some of our large white pines did survive. Out of curiosity, I counted the rings on two or three of the largest white pine stumps—they were between 125 and 135 years old. I suppose they were seedlings growing up in the wake of the logging of the original pine forests. The pines grew so prolifically in this region until the land was stripped of every last tree in the logging frenzy of the 1800s. Only the moss-covered mounds which were their stumps remain to hint at the grandeur of the original forest.

As tornadoes go, this one was a nonevent, not even newsworthy enough to be mentioned other than in the local newspaper. The damage was not inflicted upon works of man, as when a storm plows through a shopping center, a town or trailer park. It basically only decimated nature's work and thus doesn't count for much in the annals of storms. Many would argue that this is part of the natural scheme of things—survival of the fittest so to speak. I can't argue that rationalization, realizing it was more a tragedy of a personal nature than one of broader consequence. But to me it was a very real loss, and it hurt. I guess as we each live out our lives, there are those things, peculiar only to the individual, that touch us most profoundly, whether for good or ill.

At the time I had no inkling of the long-term environmental consequences to be brought about by the storm. Over the next year or two, changes in vegetation began to become very evident. Small deciduous bushes and trees such as thimbleberry, dogwood, and mountain maple took over and became thickets. Small coniferous trees, such as balsam and spruce, began poking up from the forest floor. The medium-sized pines and cedars, whose foliage had been sparse and wan under the canopy of shade, took on color and rejuvenated growth. A scraggly white pine, which had been fighting for its life amidst some cedars, has grown to have ten times the foliage it had BEFORE. Its grotesque shape, predetermined in its previous quest for sunlight, has given this now verdant tree a unique and interesting appearance to be enjoyed every day from the kitchen window.

Trees weakened by the tornado and no longer buffered by other trees from the wind, have fallen in subsequent storms of far-less intensity. Others were not able to accommodate themselves to the change from shade to the heat of direct sunlight. Wildflowers began to pop up where none had been before. The top two-thirds of the tall ash tree, which had been completely stripped of all branches, never recovered. Yet, not 20 feet above the ground, new shoots sprang out of the large trunk and have become, in this short span of time, branches as large as my wrist and at least ten feet long.

Within the month, Greg, the landscaper, and I had identified "lanes" through the trees into which he could back a large tree spade to begin a randomly patterned replanting of large (18 to 25 feet) white pines, spruce, cedars and red pines. In the area behind the house, he developed a comprehensive landscape plan for the planting of trees. At my insistence they were all to be as large as could be planted with the biggest tree spade. Greg argued vociferously that he would rather not transplant such large trees as the shock more often than not sets the growth back several years, and the incidence of failure is quite high. He was guaranteeing them. Also, he suggested that varying the tree sizes would create more interest. After some consideration, I returned to my basic premise of recreating an instant forest. I'd take my chances. That autumn he planted thirty trees, all in excess of 18 feet high. The difference those thirty trees made was breathtaking when we first saw them in place.

Once with the program, Greg's aesthetic sense, combined with his love of trees and his enthusiasm, began to create a forest landscape filled with delightful vignettes; an artful combination of the new with the indigenous. Of those first thirty coniferous trees planted, we only lost one red pine and one cedar. The others just took off from the start. By now some are reaching above 35 feet in height. Their differing growth rates have created the visual interest of variety. It must be noted that they were not simply planted and left to their own devices. They have been constantly looked after, watered in times of drought, fertilized, talked to, faithfully caressed, and had the doctor called in to investigate, diagnose and remedy maladies, real and imagined. I am so glad I stuck to my insistence in planting the "instant forest" rather than being sold the idea of plant 'em smaller 'cause they'll grow faster. And these thirty were just the nucleus.

The survival of the large red oak near the house was a real question mark. When springtime came, we arrived each time with great trepidation looking for signs of blossoming. The cold lakeshore microclimate kept us in suspense until the beginning of June. The oak was the last of the trees to bloom and, when it did, it produced sickly, yellowish leaves. "Oh no!" A panicky call to Greg brought him out to reassure us that sickly yellow was the normal color of the emergent foliage, and that it would rapidly turn green... he thought. It did, within a day or so, much to my relief. Within a month the leaves became as big as dishes, producing within me wondrous excitement and joy. Such events (and they are always occurring in one manner or another) began to provide me with an education and an awareness of the life systems and problems of the various species we were dealing with; and consequently, a far greater appreciation of these living things.

# Chapter 8
# Taking Shape

By early July John and his carpenters had the roof decking on and the building sheathed. There was a storm of activity within and around the house. Not only the carpenters, but also painters priming trim boards on horses in front of the house. Roger and his helper were installing the electric furnace in the crawl space and the duct work runs to the registers. When they weren't working on the heating system, they were roughing in the electrical and plumbing. At this time it occurred to me that I had forgotten to include an air return high in the vaulted area of the central house. There was a low return in the floor, located under the stairway to the loft, but this was not going to do the job with the heat rising and pocketing in the truss area under the ridge 18 feet overhead. I have felt this sinking feeling many times before on projects when we have discovered that something was done wrong, or had been inadvertently omitted, and the construction progress was beyond where it was doable or easily correctable. In this case, it appeared that every square inch of space (and I was looking for 240 square inches) was used. No spare space was available within a wall without destroying the structural integrity, nor up the roof slope within an attic space. Now what? Lunch on the beach was not very restful that day until I remembered there was a void between the bookcase next to the living area's fireplace and the bedroom fireplace.

"I'll be right back."

I grabbed a carpenter's rule and Roger. The space was 12 inches by 27 inches, and it appeared that the duct could be snaked up through the unheated attic and properly insulated. He thought it would work. The balance of lunch was much more relaxed. But what if I hadn't been there that day or recognized the omission until the house was much further along? I loathe those propeller ceiling fans.

Another time my coincidental presence on-site forestalled what could have been an expensive mistake by the masons. We happened to be on the scene the day they began laying up the stone fireplaces and chimney columns. Selections of variegated stone were laid out on the floor and were being incorporated into the masonry coursing on the wall face. They were just about to set the single-piece hearthstones. The one to be used with the larger fireplace took eight men to carry. But something didn't look right. They had already laid up several feet of fireplace stone above the level of the hearth, which indicated they had not followed the drawings closely enough. They hadn't visualized that the hearths were to be set in place first and the stone facing above laid on top to avoid the need for any cutting of the hearthstones. When it was brought to their attention, the masons realized their mistake and were intuitively aware that I was not going to buy into it. With good humor, they dismantled their morning's work, thankful the mortar was uncured, then set the hearthstones in place. By the end of the day, the stonework was well along and everything had been forgotten.

I wonder if the masons fully realized their artistry in stonework. There is a world of difference in laying up stone than doing brick masonry. The individual stones varied between one to eight inches in height and a half a foot to four feet in length. The face of each stone is unique in character. The selection process for texture, and the eye for fitting, is a contemplative one of patience and a sense of the aesthetic. It takes time, care and pride of craftsmanship. When we arrived the next time, the masons were wrapping up the exterior chimney stonework. Before I could open the car door, they were on the way down the scaffolding to show me the interior finished product. They stood beside me, as serious as two men could be awaiting a trial judge's verdict. The beauty of their work was readily apparent, but it took a few minutes of registration upon the detail to comprehend the incredible mastery they had achieved in their medium. Their countenances changed from worry and concern to grins as broad as happy-face caricatures as our compliments reached their ears. Those blessings of an ecstatic owner, whose reputation for perfection had been tested, were voluble affirmation of their sense of satisfaction with their work. It can be a real "downer" when you feel you have a winner, and the client either can't visualize it, or isn't of a like mind.

Early on, Jan and I were talking with Don outside the location of the back door. It dawned on us that the doorsill was much lower than grade. Don said he had followed the elevations and grades shown on the site plan to the inch—and then was mute, smug with the knowledge that it wasn't his error. Let the architect solve the problem. I had a two-foot grade elevation problem, and we couldn't regrade back farther from the wall of the house, as there were large trees within six feet of the wall. In my site layout, I had used the contour lines depicted on the land survey. They were not finitely accurate in location on the small-scale survey drawing. No one's fault really, just one of those things. The only solution which popped up was the construction of a low retaining wall to separate the walk area to the doorway from the higher natural grade beyond. I asked if he could do it in stone. He replied that not only could he do it, but he had just the right stone. We staked the outline of the wall, and he said it would be done by the next time I was up.

Ten days later we were looking at an absolute masterpiece of dry-laid, weathered limestone retaining wall. The color of the stone was exquisite; the rugged, weatheredge cream, mottled with charcoal and green with moss. Don was at the job bursting with well-founded pride at his beautiful craftsmanship, as well he should have been. I honestly think he was there only to receive the accolades for his artistry. I asked him if the railroad tie retaining wall shown on the drawings just west of the house was going to match the stone in beauty. He said he had acquired sufficient railroad ties and was going to start on it. The angular railroad tie retaining wall was cut and fit to perfection. This in its place has turned out to be a soft and unobtrusive background for the surrounding vegetation. The junipers hang over the top edge, and it is fronted by a large dense clump of indigenous yews reaching from the oak tree, down the swale, to the wall.

As summer was winding down also was the construction activity. The cabinetry had been installed with its maple butcher block tops, a soft hue against the cedar's darker patina. The prepainted rough-sawn paneling gave the central house much of its finished appearance. Having removed their table saws, the carpenters were shoveling and sweeping out the carpet of wood chips and sawdust. Inside the house, the damp odor of drying plaster permeated everywhere, overpowering the normally pervasive aroma of the cedar.

\* \* \*

After consulting with John, a move-in date was set for the fifth of October. *Aurora* was to be decommissioned a couple of weeks early to free up the time for the Schloss—as the house had begun to be called by Jan and me. Two years before we had visited Rotenburg ob der Tauber, Germany. A rare and fascinating medieval walled-town, surviving and thriving in the late twentieth century; an example of living history. The town sits on a bluff overlooking the Tauber River Valley, which was ablaze with autumn color. Visible along the meandering river was a grist mill and a small fortified tower that reminded me of a rook chess piece, except it was typically Germanic in design with the upper story constructed of half timbering. The guidebook called it the "Summerschloss" (or literally summer castle). At some point in the Middle Ages, this dwelling was a gift from the townsfolk to their mayor in appreciation for his saving Rotenburg from invasion and pillage by the eastern hordes. If I recollect, the legend is that it became a challenge of who could drink who under the table, and the good and valiant burger had no peer when it came to quaffing a brew—not even a fearsome Tartar. So they built him a summer cottage of sorts; a three-room house with one room on top of the other in the fashion of the day. Kitchen and service were at entry level, living above and sleeping on top. Enjoyable spending a half-hour snooping through this medieval second home.

Along the Drive many of the houses and even properties ("If and When") are identified by signs of varied type and design, most in keeping with the rustic nature of the surroundings. Some very clever and others prosaic. I had never wanted any such cutesy identification, leaning always toward an understatement of fact or purpose. Jan mentioned that she thought we ought to call the house Winterschloss; a twist on her recollection of the Summerschloss beside the Tauber River. It had the same fond connection within me as it did with her, and was certainly appropriate as the off-season was when we anticipated the house would get its most frequent use, due to the ever present interference of sailing. It just wasn't intended that it be so advertised. By the time construction was complete, reference to the Winterschloss had contracted to "The Schloss" and had stuck. It was then I got the idea of the sign—Schloss in German script, with a small, directional arrow pointing up the driveway, on a cedar board with the background carved away. The signboard itself was to project from the side of a carved post about three-and-one-half-feet-high. The design was given to Dan, the cabinetmaker/guitar builder, to render in rough-sawn cedar with the raised script painted white. He also volunteered to install it.

With the bedroom wings plastered (the walls in a skip coat and the ceilings in a soft-dabbed finish), I felt their appearance was rather stark and asked John to install beams across both ceilings, purely for aesthetics. The rough-sawn cedar beams not only weren't holding up anything but needed to be held up themselves. So we followed through with the bolts and black, iron plate connections visible throughout the central house. The "add" of these cinnamon-colored beams lent a great deal of warmth to these spaces. In the master bedroom, the color was a reflection of similar hues in many of the stones in the fireplace wall. How often on jobs I have seen things after the fact I would have liked to suggest as a "neat touch" but haven't with many clients, because additive costs drive some people to distraction, where others are open-minded to suggestions which might give a space a special look and end up becoming a delight forever. This more often than not has nothing whatsoever to do with the client being able to afford the EXTRA. In our case this and the stone wall were about the only significant adds, and both were generated by the architect.

I hired Ted, the Palmer Johnson painter, whose skills with a brush made *Aurora* look beautiful each winter, to do the prepainting of the interior panel boards on nights and weekends in his garage. This done, Ted was to dip all the cedar shakes and cedar shingles in clear, wood preservative. He didn't get at the job soon enough. The golf course beckoned on the long, summer evenings. When he did begin, he realized he couldn't keep up with the carpenters. Ted had a tank set up on horses on the beach and, using rubber gloves, he immersed each shake and shingle fully and set them out to dry. I figured it would take about $300 worth of wood preservative, so it was with that sinking feeling that I'd take the call from Ted every few days letting me know he was running out of wood preservative once again… and this time he was going to go to Green Bay to get it in five-gallon containers. The cedar shakes and shingles soaked the stuff up like a sponge. The idea was worthwhile, but in the end it cost over $2500 in materials and endless labor. Once into it, I just had to hang on for the ride. I didn't learn until much later, when I was about to embark upon another such ride, that it was lucky the operation had been accomplished out on the beachfront, because the vapors from the wood preservative were very harmful to vegetation (such as overhanging foliage of trees).

Ted really was a busy guy in between golf games. When the dipping was complete, he had the dining table to stain and wax–four coats of wax. At least he didn't have to sand in between coats as it would have been with varnish. By the time he had completed this job, the interior of the house had been cleaned of construction materials and debris. The loft deck, which had been covered and conveniently used for storage, now saw the light of day. With dismay, I realized I had forgotten to have the deck finished–out of sight, out of mind. This became Ted's next task. Power sanding the cedar planking got the interior just as dusty as it had been prior to the cleanup, so this entailed another cleanup, this one by Ted. Two coats of satin varnish brought out all the depth in the cedar's natural color. To this day it's never been refinished.

Window coverings had become a real dilemma. Neither of us wanted curtains, nor drapery of any sort, yet something was essential to preclude living in a fishbowl and to minimize fading when we were not at the Schloss, which was the majority of the time. A friend's house provided a creative solution, and the only one both of us could agree upon–white canvas awnings which rolled up with a clever system of pulleys and pull cord for each opening. The cord was fastened to a cleat at the base of each sash. They make no statement whatsoever, figuratively disappearing when rolled up, yet are there for privacy for a bath or sun control. The upper half of the two-tier bay facing the lake has no window covering. Day in, day out, the sunlight floods the interior space with light and solar warmth, as previously touched upon. Custom-made canvas window coverings don't sound like much, and they were not expensive, even sixteen of them with the hardware. It may be difficult for the reader to visualize, but we felt this was the ideal design solution for the Schloss–practical, durable, nonfading and aesthetically neutral.

# Chapter 9
# Move-In

The anticipation of moving in was getting to me. I was counting the days. Because of other commitments during the final four weeks of construction I only saw the Schloss once and Jan saw it not at all. This wasn't terribly troubling from the standpoint of monitoring the work. Over the course of time, I had grown to have a great deal of faith in John and his understanding of how I wanted things. He often called me from the job with questions about details or other concerns, "how do you want"… "I think we've got a problem…," or "how 'bout if we do it this way"… "I'm not sure, so we'll wait until you come up…" I never had any unpleasant surprises and for someone who never wanted to undertake a homebuilding venture because I dreaded the hassles and the resultant loss of sleep, this owner/architect-contractor relationship had grown along the way to become very harmonious and without disappointment in the charge the contractors had been given. Lucky me!

The weekend we decommissioned *Aurora* provided the singular opportunity to visit the house. And this only at night with one of the crew as company. We brought a flashlight to find the hidden key. The carpenters were just beginning the oak flooring, and the bathroom ceramic tile was close to completion. One week down the track was move-in, and there was a lot yet to be completed: interior trim painting, sheet vinyl flooring, baseboards, door and window hardware and a myriad of other miscellaneous carpentry items.

Roger had not received any of the bath hardware, and I was not willing to wait, so I told him I'd get it all in Chicago, and we'd deduct it from his plumbing contract. He seemed relieved. It took a lot of phone calls searching for things in stock, but over the next few days I had everything except the Kohler shower-curtain rods. Getting these out of Kohler was like pulling teeth. The dealers didn't have them in stock, but the factory did. I said I'd get them from the factory in Sheboygan on the way up. "Can't do that because they won't sell directly to you." Making these arrangements almost took an act of

Congress. The factory wouldn't let me pick them up without a dealer authorization. The dealers wouldn't sell them to me without the rods being shipped from Kohler to the dealer. I was getting nowhere until I finally reached the owner of one of the plumbing supply houses who volunteered to call a vice president at Kohler and arrange the unusual direct customer pickup at the plant. I put quite a lot of mileage on the car that week rounding up the miscellaneous missing items which would enable us to set up housekeeping.

My friend with the house in Door County volunteered the use of one of his trucks and crews (he's in the office furniture business) to pick up all the pieces at the furniture store and then load everything stored at our house. Then Bill and Sue would drive the truck north and help us unload.

The wife of PJ's carpenter foreman had upholstered all the window seat cushions and some pillows. Barbara and Dennis volunteered to deliver the cushions and to help unload. Ted, the painter, said he and his helper would bring the dining table out and give us a hand. It poured all day until we were within ten miles of the Schloss. Our car's backseat was packed so tightly to the roof that it was difficult to shoehorn in the shower rods. The front seat was even stuffed between the driver and passenger. Muggs rode up on Jan's lap.

We arrived at the house on a very soggy evening. Jan, who hadn't seen anything for four weeks was rendered silent in awe at the change in the landscape provided by the tree planting and the winding stone walk. Greg and his crew had been busy in the past week or two getting all the major landscape work done. Seeing the house complete (or almost complete) in its natural setting and then stepping inside to view all that had been only visualized in her mind's eye, changed Jan's speechless admiration to tears of gladness at this moment of realization. For the past year, our activities had been aimed at reaching this event. I had nonetheless emotion, but I also had four people there who wanted to get on with it. When we entered the house, a vacuum cleaner was screeching in the master bedroom. Ted had brought one of PJ's industrial vacuums to leave for the weekend and was working on the carpet. Five of us unloaded the car and truck in what must have been record time trying to keep things from getting soaked in the recurrent drizzle. The central house filled up. The multitude of boxes were left to be unpacked and organized in the morning with the exception of our bed, which we set up that night. All the volunteers took off for less arduous endeavors of a Friday night after having a celebration toast. The bottle and glasses had been carefully stowed in a box for ease of access to commemorate the launching of the Schloss.

We had two days in which to organize everything, and to prepare a punchlist for the trades. John had said he would be back on Monday to remain as long as necessary (days, not hours) to wrap everything up to our satisfaction. We didn't learn until Monday that it had been 5:30 p.m. on move-in day when the carpet and sheet vinyl work had been completed. We arrived at 6 p.m. That's what you'd call cutting it close. John commented that these tardy installations had given him some anxious moments the prior week. I'm glad I hadn't been aware of this dilatory response to the deadline by this subcontractor. I know I'm cynical, but this is so typical. They put off things until they are threatened with death or castration. Arriving at the eleventh hour to do a two-hour job, they end up not finishing, because a glitch occurs and it takes four hours. The experience of past glitches never seems to be a factor in subs' planning. At least with nonunion craftsman, they may stay beyond the bell to finish.

What a high it was to awaken the next morning to a cloudless sky and gaze out over the blue water shimmering in the early morning breeze. No window coverings had been installed; they came up in the truck. To accommodate the necessity of eating, we set the gateleg table in the small bay—the only free floor space. In my enthusiasm to get on with the day's labor of love, I wolfed breakfast even faster than I normally do. But not so fast that I didn't appreciate the exquisite view of autumn's changing colors through the branches of the venerable oak outside the window. Jan unpacked, washed and stored everything for the kitchen according to her preplanned arrangement. But only after washing the cabinets and drawers to get rid of finger marks and the film of dust. And I, beginning in the master bedroom, washed, cleaned and vacuumed every square inch of surface area in every room including all the trusses. If not then, when? Two neatniks in paradise. By Sunday night everything was basically in place and protectively covered. The dumpsters were filled beyond capacity with all the detritus of move-in. After two twelve-hour days of work-filled bliss, bedtime meant instant sleep.

On Monday morning we awakened to John's familiar voice saying,

"Good Morning." He and the painters had arrived just after 7 a.m. The painter's work had nothing to do with the nitty-gritty wrap-up. They were far from finished, both outside and in. They had only primed the interior trim work.

The gable at the end of the attic over the master bedroom had an opening which was to be closed with a small, louvered door. This door was at the painter's shop being finished, and the opening was closed up with taped construction plastic. I suggested that if a strong wind came from the east the plastic would be torn off; this to give some urgency to getting the louvered door installed. Sure enough, about 2 a.m. on Tuesday an onshore gale began out of the southeast. The noise from the surf was such that normal conversation was difficult. By about 3 a.m. the wind had shifted sufficiently that the rain was pelting against the bedroom's east window. I began to have some concern about the opening into the attic above my head. If the plastic closure came loose from its lashings, the rain would have direct access into the attic to quickly find its way to the top of the plaster ceiling. Sooner than later it would be dripping upon my head through water-soaked plaster. A new ceiling, shot!

There was going to be no more sleep this night. As I had no ability to get high enough on the wall to rectify the problem, about 4 a.m. I decided to call John and appraise him of the imminent disaster above my head. He said he'd be down. Over an hour later the painting contractor showed up. John was no dummy. By the time he arrived, the plastic film was flailing around in the gale like a torn sail. He installed the louvered door which had been completed but never brought to the job. He was a very tired painter on the job that day.

On Wednesday afternoon when I flushed the toilet in the master bath, I noticed it was reluctant to empty. Its next use filled the bowl to the brim on the flush. It's such a helpless feeling watching the "water" rise in the bowl, wondering if it's going to overflow and how much! To determine where the blockage was in the system, I flushed the toilet in the other bath, and it filled to the brim but didn't overflow. I figured I had that one safe flush.

Now what? Call Roger, who of course wasn't home. I left the message with his wife, who was somewhat less than sympathetic. They were going to a family birthday party. In retrospect, this was fortunate as Roger would be getting home on time. He did call at 5 p.m., and said he would come right down. The system had to be routed at the cleanout in the crawl space where the branches joined. This posed a real problem as opening up the cleanout would cause all the trapped effluent to pour out into the crawl space. He emptied the piping bucket by bucket with me carrying the contents up from the crawl space, out the door and into the woods. He knelt in front of the cleanout with the bucket between his legs, carefully, but not carefully enough, opening the cleanout time and again. The front of his pants looked and smelled like he'd fallen into a privy. Feeling sorry for him, Jan asked if he'd like to share our quicky dinner. Something must have told him he wasn't going to get to the birthday party, so he took her offer and sat in the hatchway to the crawl space eating pizza pie. Having routed to the full length of his equipment (40 feet) with no success, he was stumped. I suggested we contact the sewer/holding tank installer. Keith said he would come down with a longer router. When this operation proved a failure, it was after 10 p.m. They decided to dig up the sewer line at the bend ten feet up from the holding tank. By the light of a flashlight held by me, the two of them dug away. Only their exertions kept them warm as the temperature had dropped close to freezing in the star-filled October night. By midnight they had uncovered the line and broken open the elbow joint to find a cedar shingle–yes, you heard me, a shingle–jammed in the bend. This has become one of the great mysteries of life as the sewer line and internal plumbing piping were finished long before shingles were even on the job. Keith closed it up and we tested the system, and Jan, thankfully, used the toilet. Keith said he would be back in the morning to install a cleanout on the sewer line at that bend. One that could be accessed from the surface of the ground.

On Saturday, everything was pretty much in place and well organized in the interior of the house, although the floors were still covered with cardboard along the traffic patterns. In the early afternoon, I drove into Sturgeon Bay to get some things done on *Aurora* and Jan said she was going to relax with a bath.

While Jan was luxuriating in her leisurely bath, she heard a woman's voice, obviously in the house. The voice was directing other voices to look at various aspects of the interior architecture and mentioned in passing, "They must be close to moving in with all the furniture in place." The tour guide was so busy listening to herself authoritatively discoursing on the house that she never heard the shouts from behind closed doors. Quickly analyzing her predicament, Jan flew out of the tub and wrapped herself in a towel before the tour could reach her. When she opened the door to confront the group, its leader was reaching for the doorknob from the other side. The reader should know that Jan is a very private woman, uncomfortable even with locker-room nudity. It was only with supreme effort at self-control and the need to hang on to the towel that Jan asked in a modulated tone what they were doing there. Six dumbfounded faces peered over the shoulders of the guide as she exclaimed, "Oh... I didn't know anyone was living here! I have been admiring the house and am always showing my friends and guests through."

Before the dripping lady could snarl "Get out!" a few of the tour group, smarter than the airheaded guide, were backpedalling away from the scene, searching for a quick exit. With great aplomb, or rather brass, the guide motormouthed out the door saying as she waved she hoped she would see Jan again with never the courtesy of even a "pardon me."

On Friday afternoon, John had finished his final chore, installing the hardware for the canvas awnings. The painters had finished inside and were back working on the exterior trim with which they would still be diddling when we left the following Tuesday. The following day the weather turned ugly, and they never did return to finish the final coat on part of the house. This became a subject discussed at final payout time.

Over the week working along with John, we had gotten to know his quiet personality and had more than a few grins listening to his wry humor. I told him I was going to miss him but was glad it was all over. My thank-you was woefully inadequate to express the gratitude I felt to both him and Don for their conscientious efforts, their promises kept, and their willing attitude to go along with my way of doing things. It had been a most happy experience—one that at least in the broad perspective kept the genie of my fear of embarking upon this "busman's holiday" in the jar.

In turn, John said that although it wasn't by any means the largest house he had ever built, it was far and away the most challenging and —music to my ears—the most precise. He also expressed that he and all the carpenters had thoroughly enjoyed the experience and were proud of the finished product.

I now have been involved with the building industry for thirty years, and I can state without any equivocation that I had never before, nor since, seen finish carpentry of the quality found in and on the Schloss. The joinery was perfection. This is stated in the past tense as with the passage of time, and seasonal changes in humidity (and no induced humidity), some shrinkage has occurred. This has certainly justified prepainting the sides of all the interior panelling boards and cabinetry.

During the night on Friday, it slowly became apparent that the furnace wasn't working. By dawn it was COLD, and I was angry. At 5:30 a.m. Roger got a call. Being polite and waiting until a more decent hour might have meant talking to the infernal machine and missing him altogether until God knows when. He showed up with his ever present companion, Sasha the Lab, by 7:30 a.m. and within fifteen minutes popped his head out of the crawl space hatch saying: "It's only a solenoid. I've got a spare in the truck." Sasha, not understanding why she could no longer come into the house with him, sat at the backdoor until Jan asked Roger if he would put her in the truck because it was driving Muggs crazy, she having assumed a totally possessive posture over her new territory.

On the final day of our eleven day move-in vacation, I was just about done with the exterior cleanup. I had been attacking a six-foot-high pile of wood scrap, culling out the cedar suitable for kindling from that which I loaded into the wheelbarrow and tossed in the dumpster. I had filled twelve cardboard cartons with kindling wood, which I had sawed to length and chopped to size. My only saw was a bow saw with huge alligator-like teeth more properly used for cutting logs to length than 1 x 6s. Cutting the last piece of hundreds proved my undoing. The blade jumped off the 1 x 6 and into my forefinger. It didn't take long to realize I had a serious problem. Wrapping the finger in a towel, I suggested that we'd better get to the Sturgeon Bay hospital quick. Jan was pale as a ghost. She doesn't fare well looking at blood, but she made it, and they stitched me up. A carpenter I will never be, always missing the nail to the agony of my thumb. In this case I did know better than to use that saw to do that job, but I was too lazy to drive into Sturgeon Bay to buy a proper saw—which I did anyway after the fact; and have over the course of time enjoyably cut and chopped a ton of kindling wood.

# Chapter 10
# The First Fall

It was to be two weeks before we would again be at the Schloss, and it seemed an eternity. This time up, Don was going to be at the property with his bulldozer to grade the driveway and fill with 12 inches of limestone screenings to become at once the base and surface course. I had no desire for an asphalt driveway—too much like suburbia. He calculated he needed at least 17 cubic yards and said he would use a large semitrailer with a bottom dumper rather than smaller truckloads. I questioned the ability of any semitrailer to negotiate the serpentine route. The plan was to back up with the full load and then dump as he was pulling out.

"No problem."

"Oh yeah!" When the driver of the stone-laden semi saw what was expected of him, he was more than dismayed. Try and try again, he couldn't contort the trailer sufficiently to even begin the backup into the driveway entrance, regardless of all the expert advice and direction coming his way. I suggested he pull into the driveway and begin his progressive dump going up the drive as the only avenue to success. God knows what he was going to do when he ended up on the beach with the cab pointed toward the lake and the need to turn 180 degrees to get back out. He did get the truck into the path of the driveway without taking any trees with him. The trees along the route were still boarded up. With his full load, he opened the hoppers and gunned it up the driveway. Something went wrong. The hoppers were open, but nothing was dropping out. Near the crest of the driveway, the heavily weighted trailer ground to a stop, mired in sand up to its axles.

Everyone stood around staring at the disaster wondering, "Now what do we do?" It unfolded that the trailer had been loaded the previous afternoon during a rain shower. The cargo of wet limestone screenings had compacted overnight into a solid mass. The only way to solve this was to shovel by hand under the hoppers and get a rake up inside to loosen the bridged stone. I'm glad it wasn't my problem.

A couple of hours later the stone started to flow, and Don hitched his bulldozer to the front end pulling the truck over the crest and out onto the beach... empty. By mid-afternoon he had graded the limestone and compacted it with the Cat sufficient to get the semi back out, but it could not turn around. There was no room without getting hooked up with the branches of the oak or alternatively getting stuck in the loose beach sand. Using the Cat, Don literally picked up the rear end of the trailer, rotating it bit by bit, until it and its cab were pointed in the opposite direction poised to renegotiate the narrow twisting route. The driver was most anxious to be away from this debacle. Many hours before, the owner of the semi had been out to see what happened to his truck. After venting his anger with a running commentary vociferously critical of his employee's intelligence, he angrily stalked off to his pickup and disappeared never to be seen again. With the driveway in place, the only thing remaining was to remove the snow fencing around the oak tree and unwrap the 4 x 4s from around the trees.

Late that October afternoon a strong onshore wind arose bringing with it a high surf, which continued into the night. After dinner, Jan, my mom and I sat on the window seat and watched a full moon rise behind the pines along the point. Its luminescence shown bright through the branches before it passed clear above the trees and began its journey across the southern sky above the seascape. We turned off all the lights to fully appreciate this grand scene and sat mesmerized watching the moonbeam catch each marching sea as it rose and broke, momentarily shimmering firelike along the expanding length of the crest. The ribbon of light snuffed out as the power of the breaking sea was absorbed by the rising beach. The tongues of water washed high on the beach, a silvery iridescent foam fading into dark, wet sand as the wash receded. The moonlight poured through the high bay window silhouetting a geometric tracery of the timber trusses and window muntins in shadowy form across the floor and furniture. At the time, I thought to myself that it would be impossible to become bored or blase looking out upon that shoreline vignette. And I never have. The magic of this show of sound and light never diminishes.

\* \* \* \*

# The First Fall

A nearby volley of bangs from guns awakened me just after dawn on a dreary November morning. When it was repeated a few minutes later Jan suggested that maybe I ought to investigate what was going on. Dragging myself out of bed, I dressed and walked the path out to the point, which was not easy because of the fallen trees from the tornado. Beyond the point, three hunters were sitting on folding chairs in a blind at the tree line close to the water's edge, blasting away at the flocks of ducks skimming along the shore.

I was on shaky ground because this property was not mine, so I politely suggested that their shooting was very disturbing so close to the house. I was informed they'd been hunting there for years, which meant, "Get Lost!" Frequent use of the blind was evident from the multitude of shell casings and pop cans strewn about. It's a difficult situation. Some hunters feel they have inalienable prior rights despite changing conditions. They perceive their favorite spots as being encroached upon by interlopers such as myself. These mentioned they were leaving anyway, so a confrontation was avoided. As I found my way back to the path, I wondered if they, in fact, had reached their blind by walking up the driveway and around the house which was located on the old track… as was their prior right. I found myself wishing the property was mine, not George's, thinking that if it wasn't hunters disturbing my morning sleep and peace of mind, it might someday be a bulldozer carving up the property and filling the swamp to plunk another house down close aboard my own. The only remotely buildable land on the adjacent property was next to my property line. It left me with the nagging thought that I ought to do something about this sooner than later. But, I was tired of spending money.

I mentioned my early morning experience to Jim, and he said, "I used to hunt out there all the time. It's one of the best spots along the Lake Michigan side of the peninsula for ducks because of the point jutting out." For starters, I called George and asked him if I could post his property in his name and have his blessing to get rid of the blind, all of which he agreed to. A weekend or two later, I found three dead ducks washed ashore on my beach–ones hunters could not retrieve. Damn, this made me angry and determined to put an end to the nearby hunting. The NO TRESPASSING – NO HUNTING signs did not stop the practice but did give me a greater posture of authority. I had occasion to test this several times with little resentment caused.

One day after hearing shots, I came upon an empty blind and continued on to reach a group of hunters who had established themselves in a hastily constructed blind directly in front of our nearest neighbor's house. They had trimmed some of the neighbor's trees for camouflage and had even parked their car in his driveway. Not knowing if they were friends of the neighbor, I didn't get into a discussion but when I next saw him I did ask about it with a predictable response.

The absence of any beach on the far side of the point was critical to hunting success. Lake Michigan crashed ashore almost at the tree line, providing the foreshortened range necessary for shotgun ammunition to reach its victims. On our side of the point, the shoreline opened out onto a rather broad beach. Three downed pines stuck out over the water making any walk around the point impossible even on a day with a calm sea. With my activities in the house reduced to tinkering and touching up, I began to take notice of our surroundings. The fall storms were causing some changes along the high-water mark at the tree and vegetation line. Stones and rocks washed clean were lying on the forest floor several feet in from the tree line. In a storm, I watched them being tossed and left high and dry as the seawater soaked into the layer of duff. The roots of trees at the shoreline embankment were being exposed from the soil being washed away. A few smaller trees leaned out at crazy angles.

It seems that each time we were up we experienced another severe blow out of the northeast. At these times sitting in the window, I could see the surf wash up the beach well beyond the first undisturbed rise of sand covered with beach grass. This was within 75 feet of the window. Even in the well-insulated house, the noise of the surf made normal conversation across a room impossible. Sleep at these times had become a real trial, as the pounding of the surf was a relentless reminder that the sea was beginning to cause significant erosion and beach loss. It was during this autumn that I began to see articles in newspapers and other periodicals concerned with the rising lake level and the damage to bluffs, beaches and shoreside facilities. The Corps of Engineers, the organization monitoring and predicting lake levels, was forecasting even higher water in 1986.

\* \* \* \*

The watercourse of the creek is well-defined until it reaches the beach and is subjected to the powerful forces of wind-borne wave action. The most direct route of the creek to the lake is perpendicular to the south-facing shoreline. Should the wind come up from the southeast, the wave action immediately begins to push the creek's mouth westerly along the beach. The stronger the wind and the longer its duration, the further down the beach the mouth of the creek moves until it may have shifted 100 yards or more. With this extended route to the lake, the creek flows parallel to the shoreline once out on the beach. The current in the creek erodes the sandy banks on the outside perimeter, accelerating its progressive movement down the beach. A contrary force of current is set up at the 90-degree bend where it comes out onto the beach. The centrifical force of the current cuts away the sand as it makes its initial bend to run parallel with the shoreline. The current continues to cut away the bank until finally it scours away the last sand between it and the lake. Thereupon again beginning a direct route of flow into the lake. A consequence of this is the drying up of the long meander down the beach. Should the wind shift diagonally from the opposite direction (southwest) the lateral movement of the outfall simply reverses itself. The continuously shifting position of the mouth creates some fantastic serpentine variables. Although the shifting is subtle, the changes over a twenty-four-hour period can be astoundingly dramatic. It's not so subtle that, by observing the flow for several minutes, you can see the current continually at work undercutting the banks, dropping the sand into the water and sloughing it along with its flow. Some of the eroded banks may be three or four feet high.

A daylong storm with a high surge can completely obliterate any evidence on the beach of yesterday's creek. When the seas from an onshore gale are washing up the beach, their forces are in action directly opposed to the outflow. They wash right over the beach between the shoreline and the creek, depositing sand higher and higher until the creek is completely dammed. The higher the storm surge up the beach, the higher the dam, causing in high-water years inundation of the creek's floodplain all the way inland to the Drive. When the mouth is dammed, the creek flows into the lake right through the sand, as it would through a filter. The trickles are visible just above the waterline, and the sand is unstable. When walking upon this saturated sand, your feet sink in instantly.

When the wind and sea abate, the creek sooner than later tops over the dam and begins its rapid cutaway of the sand until the water is rushing through a progressively deepening and widening gap down into Lake Michigan. Within several hours the water level in the creek can drop six to eight inches.

The water in the creek is rusty brown from the tannin it transports from the cedar swamps at its source and along its course. On a calm day, the creek flushes a lengthy bloom of brown-colored water into the lake. When the water isolated in an oxbow cutoff filters down into the sand, drying out the cutoff, it plasters the depression with a dense surface coating of the brown particulate matter that had been suspended in the creek water.

It's interesting to think that identical forces to these are at work shaping and reshaping the largest rivers on earth. It's wonderful to have at hand this natural laboratory; a dynamic example to study and learn about these interactions driving perpetual change along this littoral.

# Chapter 11
# The Storm/Armoring The Shore

My increasing awareness of the consequences to my property caused by the high lake level prompted ever more frequent concern and anxiety. Each time we arrived I would immediately walk out on the beach to see what had occurred in our absence. When storm surges dammed up the creek, its level rose to the point where it backed up into the oxbow cutoff (the slough) reaching toward the house. There were good-sized trees in this wet swale whose health were being affected by frequent and lengthy emersion. These trees were clinging to saturated soil, sure candidates for uprooting during storms.

A time or two in very bad storms, the seas surging up the beach actually topped over the initial grassy rise and flowed down its back side toward the slough. With even higher lake levels being forecast, this had the ultimate possibility of causing a major reorientation of the creek through the old oxbow cutoff, bringing Lake Michigan far inland from the present high-water line and this shoreline interaction dangerously close to the house. Having just spent my wad constructing the house (it was only at the end of October that Don and John had received their final payout), I was in no frame of mind, nor flush with cash, to begin what I knew to be another major construction endeavor.

Installation of shoreline protection does not come cheap if it is done adequately and properly. If it isn't done properly, you might as well pour your money into the lake. The forces of the sea from which you are buttressing the land are enormous and will destroy any ill-conceived, poorly engineered solution. And that's about all I knew of the subject. A sandy shoreline contributes significantly to this capacity for destruction, as I found out when I decided it would be prudent to educate myself on the subject.

Early in November I contacted an engineering firm experienced in the design of shoreline works. Their lengthy consultation was enlightening and scary. I subsequently contacted two local contractors doing this sort of thing. One responded instantly, the other never did. Rick and I met on-site and discussed the feasible options, and I ended up with a huge headache. It became clear there were only two solutions which would not permanently destroy the aesthetics of the beachfront. The first was the placement of dimensioned limestone, quarried locally. These "stones" are somewhat regularly shaped rocks, flat top and bottom and more or less on the face. After being dynamited out of the quarry wall, they are selected for having these characteristics and for weight (being upwards of 2000 pounds). A trench is excavated along the high-water mark of the beach. Filter cloth is placed continuously along the bottom and back side of the trench to prevent washout from behind the limestone wall. The rock is set in place in layers like a giant-sized brick wall laid dry; dependent upon their own weight to stay in place. Finally, melon-sized rock is dumped in between the filter cloth and the back side of the dimensioned stone. This is a big operation necessitating trucking the rock out onto the beach and the deft operation of a large shovel that picks up each rock and sets it rather precisely in place three layers high in the pre-dug trench (an average size stone might be 2 feet high by 6 feet long by 3 feet in depth). This has great limitations on a sandy shore because excavation of the trench much below water level is impossible. Over time there is a strong likelihood the wave action will undermine the sandy foundation, beneath the bottom course of rock, destabilizing the integrity of the wall; the rock to disappear in the sand at worst, or at least to shift, requiring maintenance.

The second and more preferable solution was to drive in a continuous wall of interlocked, anchored steel sheet piling—at a cost of twice that of placing dimensioned stone. It didn't take a Ph.D. in engineering to figure out that given these shoreline conditions, sheet piling was the only reasonably permanent answer to keeping Lake Michigan at bay.

These operations had far more propensity to damage the property than the construction activities associated with the house. Large, tracked equipment had to reach the beach by using the driveway: bulldozers, air compressors, a long boomed crane, backhoes and the endless trucks bringing stone (or alternatively 40-foot lengths of sheet piling). The new driveway was going to be destroyed. Its six-to-twelve inches of crushed limestone base was simply not capable of sustaining the impact of this type use, not to mention the damage to trees, old and new, along the twisty thoroughfare. As tough as it was to accommodate myself to this disruption, I knew deep down it was going to have to be done to preclude the possibility of significant land loss to storm generated erosion. Also, if I was going to do it, it made sense to get underway as soon as possible. There was no percentage in procrastinating. I arranged with Rick to stake the line of the wall with me over Thanksgiving weekend. The decision of which method to use was to be postponed until soil borings could be taken to ascertain the depth of bedrock beneath the surface of the beach.

Rick agreed to protect all the trees along the driveway, erect snow fencing around the oak tree and isolate the entire beachfront with fencing, except in the specific work area. I said I would prepare a drawing for state approval and specifications on which to base our contract.

Working on the beach in a knifelike wind and continuous rain, we sighted the line, drove the stakes and determined the intended elevation of the top of the wall in record time. While doing this we both noted the effect the northeast gales were having on the shoreline just westward of the point. The northeast storm seas, though not coming directly ashore because of the projecting point, refracted around the point and swept obliquely toward the beach. They washed up on the beach diagonally and dissipated their energy by scouring away the sand surface and undercutting the high-water embankment. Even though the northeasterly sea was lazy that afternoon, the swirling motion was there in sufficient strength to be perceptibly eroding away the beach.

By evening the rain was turning to sleet on an increasing wind. Bedtime saw a heavy, wet snow beginning to accumulate. During the night the wind increased to gale force and the noise from the surf was deafening. Morning dawned and with it came a view of a drastically different outdoor world. The front wall of the house was plastered with thick snow. Drifts were well above the windowsills. The branches of the coniferous trees were bent to the ground, carrying a burdensome load of heavy snow. Some trees had doubled over, yet somehow not split. The wind increased in strength and the snow in intensity all morning.

I was so concerned over the newly planted trees that I spliced a long pole and rake together to pull the snow off the branches, which were plastered together like frosting on a tree cake. I don't know where it was wetter, inside my clothes or out. Soaked from the exertion of wading through knee-deep snow and maneuvering the rake above me, continually bombarded with loosened snow, I kept at my job for several hours until I couldn't lift the pole any more. When I came inside for a shower and dry clothes, the TV was tuned to the Green Bay Packers game. If I remember, they were playing the Chicago Bears at Lambeau Field, an outdoor stadium in Green Bay. It was snowing so hard in Green Bay that the screen was nothing but a blur of white with an occasional and momentary view of shadows slipping and sliding and bumping around the invisible field for the enjoyment of those few hardy fans loony enough to sit through a blizzard. So who's loonier, the football fans or the guy out in the forest raking the snow off his trees?

The TV weather forecast called for the snow to diminish by nighttime with a wind shift into the northwest pushing arctic-like temperatures into the region by morning. We had intended to leave that day, but 24 to 30 inches of snow shut down all of northeastern Wisconsin. Our snowplower couldn't do his thing because the road wasn't plowed until very late in the day, and then only an initial swipe. Jan's folks were with us from Florida, and having enough food it was prudent to just stay put. As fast as the county plows were clearing roads, the wind was closing them back up with drifting snow. When we did leave a day later, reaching the main highway ten miles distant almost proved impossible in whiteout conditions of blowing snow. I don't mind driving in snow, but this proved to be the worst I ever recollect. The expressway from Green Bay south was covered with ice several inches deep, off of which the snow had been plowed, more or less. The ruts in the ice controlled steering and threw the car about, jostling the occupants like passengers in a stagecoach. My passengers in the rear seat were either remarkably calm or scared speechless.

This has gotten off course somewhat, but it does describe the ferocity of the storm. The day following the storm my first thought was to walk down to the beach to see what happened to the shoreline. A stroll it wasn't, having to slash and highstep through waist-deep, drifted snow to stand upon the edge of the high-water embankment. Nature's latest handiwork was painfully obvious.

It had only been a few hours since the bottom fell out of the thermometer, so the sea was just beginning to armor the shoreline with ice. The beach, though ice-coated, was free of snow, where the surge riding up the sloping sand had kept it clear. I looked in vain for the stakes marking where the wall was to have been. Not only were the stakes gone, but also up to 30 feet of property. The sea had been particularly voracious in its appetite for dry land underneath the point, where it had carved away a big bowl. North of the point, less land had been lost because of tree coverage, but many trees layed horizontal, their tops in the water, having lost their root-zone support.

This was terribly discouraging and gave me cause to consider that maybe I wasn't as smart as I thought purchasing shoreside property, and building a home on the beach. Conversely, it sure gave impetus and proof positive to the wisdom of getting on with the shoreline protection work. The reader can, by now, imagine that I was burning up the phone lines to get Rick in motion, and even more important, to find out how I could expedite the permit process in the state's Department of Natural Resources, a typically ponderous bureaucracy. I ended up going to its regional office in Green Bay to pick it up.

Rick suggested that during this permit interim he plow all the insulating snow off the driveway (the truck route to the beach) and staging area. This was to permit the frost to drive deep into the ground to create in effect a paved, operating surface. This saved the driveway from any damage whatsoever.

December continued bitter cold with mountainous ice forming along the shore. Subsequent storms sent water crashing over the face of the ice to freeze solid on top, ever thickening the armor for the duration of the winter. The ice platform reached almost up to the line of the intended wall.

By our next time up (in mid-December), the trees along the driveway had been rigged with 4 x 4s (deja vu), and a temporary electrical service was strung from the road through the forest to the beach. The soil-boring rig had done its investigation, and the information gleaned settled the method of protection. At the creek, bedrock was 14 feet beneath surface. It sloped upward until at the point 280 feet eastward it was nearly at the surface. Steel sheet piling was not suitable for use where bedrock was close to the surface so we ended up setting three courses of dimensioned stone until a sufficient depth of bedrock permitted using sheet piling for the balance of the way across the frontage. Where the two met, the steel was cupped around the end of the rock wall to lock it in place. This combination required drawing revisions, but we decided to wing it without the agony of re-permitting.

Having moved into our dream house less than three months

before, we were looking forward to spending the week between Christmas and the New Year at the Schloss. I never dreamed that our first vacation along the wild winter shore of Lake Michigan would be punctuated by the "thunk" of a pile driver doing its thing all day long. The view of the winter shoreline out the bay window was interrupted by a collection of heavy equipment. The workday began before sunrise with the crews trying (sometimes in vain) to get the diesel engines to turn over. They normally were hooked up to block heaters overnight, but when it dropped under 10°F nothing did much good (and several nights it was below -10 degrees). The cold temperatures had contrarily some benefit. Beside the ground surface being frozen as solid as concrete, the shoreline ice had reached a height such that the tracked equipment used it to operate from.

The view out the window was not one you'd submit to if you had paid for a room in a four-star hotel in Aspen. It looked and sounded like a superhighway was under construction 75 feet from the window. Of course we hadn't paid for a winter vacation at a luxurious inn; we'd just bought the hotel.

It was interesting to watch the operation day by day as work progressed along the beachfront. Between this and the daily runs of cross-country skiing, the time flew along rapidly. Slowly the piles of staged rock and steel diminished.

The eastern portion of the wall, which involved trenching the beach down to the cobbles on top of bedrock and placement of the stone, was completed relatively quickly as it was only a 70-foot run. Driving the sheet piling seemed to take forever with an optimum run per day of 30 feet. With problems (and the bitterly cold weather caused many, mostly due to equipment failure in the arctic-like cold) progress was often reduced to a few sheets driven per day. If it wasn't the pile driver, it was the shovel that was recalcitrant. Some days they were still trying to get the engines running until well after 10 a.m., and some days they simply gave up until the temperature warmed up another day.

The piling as well as the stone placement required trenching. The sheets of steel were driven downward into the earth at the bottom of the trench until they struck bedrock. We could always tell when another sheet reached refusal as the sound was entirely different than hammering through sand.

The men working in the trench were lucky, they were out of the wind. Several things occurred in the trench. The individual sheets of 18-inch-wide corrugated steel were kept in line, true with the run of the wall by a horizontal steel channel welded on the back side two feet down from the top. The top of the sheets were burned off and the rim was capped with a steel angle welded in place. Upon completion of the sheet piling, the shovel went back to work digging a second trench parallel and 20 feet inland from the sheet piling, connecting it with the wall by offshoot trenches (like steps on an extension ladder on the ground). They then constructed a backup or anchor wall of reinforced concrete in the second trench. This wall was connected to the sheet piling every 20 feet by steel tie-rods. Now the redi-mix trucks were running up the driveway! When we arrived (we always seemed to be arriving) at 10 p.m. on a Friday night in mid-January, there was a tent running the entire length of the anchor wall with large blast heaters pouring, or I should say "roaring," heat into the trench to keep the newly poured concrete from freezing. The only description of that night I can think of would be that it was like trying to sleep next to a jet engine at full throttle.

By the first of February, they had backfilled the trenches and graded off as best they could under the winter conditions. All the equipment was gone and tree and beach protection removed. The scars of winter construction were soon buried deep under new snow. Rick had agreed to come back in the spring when the snow was gone and the frost was out of the ground to do the final grading and restore the appearance of the beach—without vegetation, which would hopefully grow back over time.

# Chapter 12
# Necessary Arrangements

Building a house in the rural areas of Door County entails applying for a fire number from the town. It's akin to a street address in urban areas. The fire number is white baked-on enamel on an embossed aluminum background painted green. The sign is about 14 inches by 6 inches and is bolted to an iron fence post, though I've seen them nailed to mailbox posts or trees. No big deal, call the town chairman, who handles this, and within a month or two you'll have the bright, new sign and post to go along with it. He even offered to pound the post into the ground and bolt on the sign which I thought was unusually gracious and helpful. It was years before I became aware that the chairman gets paid to do this.

Because of the unusual junction of the Drive with the access lane and my driveway, its placement was not going to be satisfactory anywhere. This was only made more confusing by the proximate location of another fire number sign; our neighbor's several hundred feet down the access lane. Rather than negotiate an appropriate placement of my sign with the chairman, I said I would install it if he would lend me the tool to drive the post into the ground, which he was agreeable to doing.

I ordered the fire number from the chairman in May as the construction was just beginning. When he hadn't called by August, I called. He said he didn't place an order with the sign manufacturer until he had a dozen or so.

"I've just placed the order. It'll be here in six weeks."

It sounded reasonable enough. At the end of September I checked in again.

"They are due any day."

In October several other homes, built long after the Schloss, had their bright, new fire number signs at the roadside. And I had gotten a complaint from the sheriff that they couldn't find the house because there was no fire number. For the life of me, at this point in time I can't remember why the sheriff was looking for the house, or me.

So I phoned the chairman again. It was as though I'd never spoken to him before. My inquiry dragged out an admission that he had never ordered the damn thing!

I politely (very politely) told him of the complaint from the sheriff. After much agonizing he said he would special-order it, and I should call back in a week or so to find out a delivery date. When this turned out to be late December, I asked if I could have the post and borrow the tool to drive it in before frozen ground would preclude this even if I did get the sign. I had visions of not having my fire number in place until the following spring. He offered to put the sign between the doors at the Schloss when it arrived.

When I got there to get the post, all he had was an old one—slightly bent. I took it anyway feeling somewhat helpless and frustrated at not being able to conclude something so routine. I'd go to the hardware store in the spring and buy a new post.

Not having heard from him, on the last day of December I called.

"I've had it a couple of weeks. Thought you were never going to call for it."

I swallowed hard and said I'd be right over to get it. When I got there it was dark, and though I could see it was the proper number, the background color didn't look right. I put it in front of a headlight, and sure 'nough it was RED. "For God's sake!" Was this fiasco never to end?

Responding to my knock, he looked at who it was and with evident dismay said, "Oh, I thought you'd left."

"The sign is the wrong color, it's red."

"All of them are going to be red from now on. Yours is the first," he glibly replied.

So I bolted on my red fire number secure in the knowledge that I was the avant-garde of future fire numbers.

The day after New Year's, a girl lost control of her car on the bend in the Drive and skidded off the snow-packed road into my signpost. It didn't harm the red sign, just mangled the post into a pretzel. So I unbolted the sign and gave up until spring when I bought a post and remounted the sign.

The unique fire number caused the inevitable questions from neighbors.

"Why did you get a red one?"

I explained to each inquirer that the town had changed its policy and all new signs were to be red from now on. I usually got a quizzical look and an, "I didn't know about that."

Their skepticism was polite, because everyone reads the local newspaper, and any newspaper that tells what the school lunches are going to be would surely have published this policy change by the town.

It wasn't far into summer when new fire number signs began appearing. Intuition became reality. I had been the victim of a bald-faced lie as all the new signs were, of course, green. But by this time the sniggering was over, and I didn't have the ambition to begin all over again. So the red fire number sign stayed until one night it disappeared, post and all. Two weeks later, I happened to catch sight of it in the woods on the other side of the Drive. So I got Greg's fence post driver and put it back in place. It stayed that way for two years until someone decided they wanted it more as a memento than I did as an address and absconded with it. It to remain, I assume, in a bedroom museum of similar purloined artifacts, all significant to the history of Door County.

Now I had to ask the town chairman for another sign, and I was going to do so with some satisfaction.

Yet I was not to have the opportunity to confront my sign nemesis' disingenuousness. In the interim, an election had been held and the town had a new chairman. He, to my relief, gave me a new, green fire number plaque within a month after talking to him, and it was put in between the doors.

\* \* \* \*

It took only a few days of use before the toilet bowls began to change color–to rust, evidence of very hard water. Never having had the experience of using well water, I was unconscious of the consequences of hard water to the appearance of the fixtures. Jan, having prior awareness of this, suggested we invest in a water softener. The suggestion fell on deaf ears until the suggestion was voiced with the emphasis of a sledgehammer. I called the "Culligan Man" who came out to price an installation. I told him there was no room to locate the equipment anywhere but in the crawl space. Instead of walking away from the challenge, he worked it out. He had to sink the brine tank beneath the level of the crawl space floor so the top could be removed to fill the tank with salt. The service man has never forgotten this installation that requires a contortionist to drop into the crawl space with the 80-pound sack on his shoulders, crawl to the softener, open the top and somehow pour the salt out of the bag held up against the underside of the floor joists. It always brings a smile (which I try to conceal) when the "bag man" disappears into the crawl space, arms fully occupied and unable to defend himself against Muggs' wet licks. It's good he's a dog person and enjoys her attention.

For several years we had endless problems with the state-of-the-art sensor controlling the frequency of softening. The Culligan dealer finally gave up on it and offered to replace the erratic sensor with an old-fashioned flow meter. This works unerringly and has reduced our consumption of water dramatically. The water costs nothing, but it all ends up in the holding tank which has to be pumped out for a price.

\* \* \* \*

Caretaking is a second job for a few of the local folks. The owners of vacation homes retain these people to watch over things while the resident is away, which is the bulk of the time. Often it seems that the absentee owner just wants the caretaker to put his sign out in front without having him expend any real effort at consistent monitoring. It's sort of like posting a security systems' ID on the doors and windows without the benefit of the alarm. I didn't want any sign, but I did want some effort and was willing to pay for it. An interview set the program, and in turn the fee, which I didn't question. It wasn't necessary for the caretaker to go inside, except when the heat needed to be turned up prior to our arrival, or when the schefflera needed watering during a prolonged absence. However, it did involve getting out of the pickup for a good look around on a daily basis. "If you don't want to do what I expect, don't take the job!" This was all I asked. Assurance of conscientiousness sealed the bargain.

Over time the caretaker's lifestyle changed, with freedom of divorce, less regular work hours, and other intrusions of a second job, if not the first. Communication was necessary, and I could never reach the caretaker to let him know of our impending arrival, a change in our plans or to water the damned plant.

"Why don't you get an answering machine?" I inquired.

"I've got one—just haven't hooked it up."

One year later it still hadn't been hooked up. I have a low-tolerance level for this sort of thing. With snow on the ground and no footprints in the snow, it was rather conclusive the job wasn't being done despite my admonishments. After having to come up to a cold house several times because we couldn't reach him, the untracked snow tipped the balance of my forbearance. I asked my snowplower and his wife if they would take over. Somewhat reluctantly (they got the plowing business from the former caretaker) they did so, the burden of the job falling mostly on the wife. I liked the original guy but just wasn't getting a fair shake in the changing agenda of his life's limited time. It seemed a natural for the snowplower.

Over time I began to feel intuitively that the service level was again deteriorating, and the trackless snow clinched it. Everyone knows everyone else in these parts, and I, by chance, learned Russ had cancer. Obviously a greater and greater burden of responsibility was falling on his wife, already busy raising two young children.

By this point in time I'd invested in a real honest-to-god security system. I was as much concerned with a power outage lasting long enough to freeze piping, as I was with vandalism or a break-in.

In bitter cold weather an outage of twelve-to-eighteen hours might just do it if the caretaker wasn't alert to the condition. There is nothing of any real value in the house, which might only aggravate a thief. We don't drink so there's no booze around, which might frustrate a vandal. The security system is simple but terribly effective. Even if an intruder managed to enter without setting off the alarm, he couldn't move two feet inside without getting zapped. It works, as one time the door was left ajar and the wind eventually blew it open. The central monitoring station called my caretaker, and the sheriff. She got there before the sheriff and was turning off the alarm when they arrived. The deputies with guns drawn scared her half to death. And they weren't very polite about it, even after she gave them my phone number to call.

Russ unfortunately died from the effects of "Agent Orange," it was surmised; somewhat like bait poisoning coyotes, killing the poisoner's dog. With all the burdens of single parenthood, his wife had to give up this peripheral caretaking responsibility. When next at PJ's I saw Dennis, the carpenter superintendent, and asked if his wife, Barbara, might like to do it. I was in luck.

If you recollect, Barbara made the window seat cushions for us. Since she was in the habit of long bike rides, she felt it would give her a reason to be doing so. She and Dennis (he gets involved with the deal at times) have been ever so conscientious. Doing things on their own volition, such as watering the plant inside, taking in the mail, watering the geraniums in the planter boxes, shoveling a path through the snow to the door. It's really been great not to have to think about the caretaking—it just gets done.

\* \* \* \*

Though I was explicit in my snowplowing instructions, Russ wasn't into finesse in this endeavor. The result being broken branches on flanking cedars and a few scarred trunks. He got the picture the second year when I had him help me drive in some reinforcing rod guides at strategic places along the serpentine route. It was back to the same old litany, "Take care of my trees, they are very special to me!"

With Russ gone, I not only had lost a caretaker, but also a snowplower who had acquired the plowing adeptness I sought. Now who? Chances are its going to require some "training." But wait, let's ask Greg, the landscaper, if he would take on this chore. He is the right guy to be doing it. He and his people are accustomed to my mindset, and they on their own revere the trees. Although he wasn't into this winter adjunctive to his business, he did agree to do it—and has ever since. A snowblower is much kinder to the trees than is the corner of a plow blade.

Friday night arrivals have always brought surprises. Russ also had done the snowplowing for the town on the access lane. I didn't find out who was going to take his place until the narrow unpaved lane was most efficiently and destructively cleared of snow by the county highway department's plow. This is big equipment and moves fast, operated by guys whose sensitivity toward flanking trees and mailboxes is nonexistent. They plow the highways ever wider onto the shoulders, the theory I suppose, to keep ahead of potential drifting. But there is no drifting on this narrow lane with the forest right alongside the pavement. We didn't want or need an expressway plow-job to service four homes.

To improve the aesthetics, I had planted twelve or thirteen large cedars and spruce along the narrow right-of-way to enhance the appearance of the lane. The plower had tried to make a winding road straight and pushed the snowbanks into the flanking trees, breaking off the lower branches. He missed the Schloss sign by an inch or so. This is like using a tractor to cultivate a flower garden. Trees are just obstacles to these people. They spend all their make-work time cutting them down to clear an optimum field of vision and take out the curves in their eternal quest for the perfect highway—justified by safety. The writer has nothing against safe highways, yet not every road must be made into a featureless expressway. There seems to be no balance in judgement or planning to maintain the aesthetic quality of roadways.

In my experience dealing with highway engineers and the departments that maintain roads, I have become firmly convinced they are environmental Neanderthals, devoid of any sensitivity to the natural surroundings. If a wetland is in the way, fill it; if it's a forest, cut it. Always the environment is sacrificed to the tune of "cost-effectiveness." The evidence of these unenlightened policies is dramatic in a place like this county—ironically one whose economic survival is dependent on folks traveling from near and far to soak up the scenery—not be entertained by a DisneyWorld.

Over the years, the county has indiscriminantly cut down the trees along the right-of-ways. Now there are only a handful of country roads where the spreading maples arch across. The tourists flock by the hundreds on busy autumn weekends to photograph these few remaining tunnels of colorful foliage.

Some places have covered bridges which people come to see. These anachronisms are certainly rickety, narrow and inefficient methods of crossing rivers in the late twentieth century. But they are recognized as assets and remain to give us pleasure. So also should the character of the byways be retained in this place.

As for my trees, the damage had been done. I did write the county highway commissioner and explained my concern, but this was no solace. To prevent additional damage before the next winter, I expected to have to deal with the problem in a more effective manner—how, I knew not.

## Chapter 13
# Winter

Although the longest I recollect, the first winter passed rapidly, carried along by our boundless enthusiasm for the Schloss. We never missed an opportunity to be in the out-of-doors. On skis, we were beginning to find our way along the narrow trails through the large, roadless wilderness stretching for seven miles just the other side of the Drive. But inside or out, it didn't matter. The important thing was just being there.

Housebound on the coldest days, I often lost track of time simply gazing out the windows at the season's silent repose, covered by an ever-thickening blanket of snow. Each window framed its own special vignette to capture my contemplative interest. From the small bay, I watched the sun set far toward the south, coloring the sky a deep pink behind the distant, shoreline icescape. The snow and ice, so white and blue in the blinding sunlight of the January afternoon, quickly fades to twilight's gray and mauve; the long shadows of the short, winter day blotted out by evening's cold monochrome.

Muggs, the dog who spends summers stretched out over an air-conditioning diffuser, was constantly enervated by the snow, scampering around, nose-diving into deep tracks of other animals for the fresh scents, never wanting to go back inside. She did remember her previous experience with deep snow so would only stand with her chest pressed against the high banks looking beyond. Night walks were a favorite despite the often bitter cold until one night when two owls were who-whooting back and forth unseen in two nearby trees. She listened thoughtfully and made an about-face for home, never stopping until she reached the back door. Only then did she recover enough of her usual bravura to stand up and peer over the snowbank, offering a low growl to intimidate whatever was lurking beyond the loom of the entry light. Since the owls, she hasn't been enthusiastic for night walks in the scary forest (or on the road, or even very far down the driveway).

Since skiing along the beachfront is regularly bone-chilling cold (as it would be across an open field), the forest behind us beckoned. We quickly discovered three or four entry points into the roadless area along its seven-mile stretch. Each time we would venture farther through the silent forest with ever-increasing stamina and a desire to see beyond the next bend. Seldom did we meet any other skiers, having the forest all to ourselves. Many times we were the first in, to break the trail through newly fallen snow. Though the wind might be strong above the trees, its only effect under the trees was an occasional movement of a branch losing its mantel of snow, dusting the powder silently downward to the forest floor, and once in a while down on our heads. Breaking trail, the only sound was the soft hiss of the skis and the lowing of the wind high in the pines. Occasionally a snowmobile would pass, racketing along the trail at breakneck speed. Experiencing the winter forest on skis can be serene and good for the spirit if the skier is not bent on setting speed records. For my money, the concentration that's required with speed dampens the aesthetic joy of just being there. "Each to his own." Within the first year or two we had ventured through frozen bogs and cedar swamps, along creeks, under thick canopied hemlock forests, over and around the ancient shoreline ridges of a former Lake Michigan to discover three small wilderness lakes. Skiing out onto these lakes from within the forest the scenes could as well have been the vastness of the Canadian wilds three or four hundred miles north.

* * * *

Small lakes freeze over and remain so until spring breakup, creating throughout winter a sense of suspended life or hibernation. The months of deep cold and snow bury much of life within. In this silent world, the aquatic life is dormant, the ducks and geese are gone, and animals aren't quenching their thirst or eating the shoots growing in the shallows.

In contrast to this seasonal dormancy, Lake Michigan throughout winter provides a kaleidoscopic drama along its shoreline. The first, cold, clear mornings bring unusual low, heavy clouds over the lake–cold air transferring the water's relative warmth into moisture close above the surface. Smoldering tendrils of a purgatorial mist evaporate as they rise from the water's surface to form a low, surreal layer of heavy, dark clouds on an otherwise blue-sky day; in my imagination, the quintessential picture of a Middle Ages hell. The seas themselves carry along a great burden of chunky ice. Sometimes so much so near shore that the ice dampens the heaving motion into very rhythmic undulations. After days of below-freezing temperatures and calm winds, the surface of the lake may be covered with pan ice stretching out beyond the horizon, held against the coastline by an onshore breeze. One day's darkness might fall upon this icebound shore, and the following dawn "presto!" not a single flow of sea ice is visible. An offshore, northwest wind worked its magic overnight to beguile its morning audience with a clear, blue sea, seen over the tops of the mountainous, ice formations clinging to the shore and grounded close offshore. This is where the real action is taking place, that which you can hear and feel, as well as see.

The change from autumn's unprotected, unfrozen shoreline to winter's icebound coastal armor happens quickly when the daytime temperatures don't rise above freezing. Even without an onshore breeze to dash the sea up the beach, a gentle swell washing ashore begins to form the thick ice just above the waterline. Once the foundation is there, a gale pounding ashore thickens this foundation hour by hour until the ice is several feet high and has advanced outward at times well over 100 feet into deeper water.

The face of the ice becomes a vertical wall against which the seas hammer relentlessly and effortlessly. It's no different than any sea wall buttressing a deep-water shoreline against the full force of inbound seas, undiminished in power by a shoaling bottom. Each wave crashes against the vertical ice wall throwing spray and solid water geyser-like, high above the top of the wall, to be blown shoreward. It all lands on the back slope of the ice hills inexorably increasing the height of the armor as it has nowhere to drain.

The skyward-bound water and spray carry with it a cargo of ice chunks which clitter-clatter down the back slope to collect on the surface. With their clarity and smoothness, they glisten in the sunlight like thousands of diamonds. Each dull boom indicates another shower of water and ice. In strong winds some of these chunks are blown far in on the beach. The surface of the lake, just beyond the ice wall, is a confusion of incoming and reflected, outgoing seas exploding as they meet head-on. Beneath the ice face, the roiling motion of the water picks up another cargo—of bottom sand and stones to dash them aloft in the spray.

Face the wind on the beach on a stormy day, if you can stand it, and soon the wind-driven spray will begin to coat you too. In the strongest storms, the freezing spray has reached the front wall of the house.

Over time, the wind-driven ice is layered higher and higher until sitting in the house we cannot see the open water beyond the continuous shoreline barrier of mounds and cones and humps. The cones are the highest; hollow inside, they resemble volcanoes in shape. The character of this shore-bound ice is anything but homogenous and, consequently, is weak and dangerous. It's subject to rapid erosion by incoming seas on warmer days. Gaps and cracks are often bridged over, traps for anything venturing over the top. An onshore gale, on a warm day, gouges deep into what yesterday seemed like an impenetrable wall 50 yards out from shore. The soft, sand-impregnated ice, many feet thick, is invaded and eaten away by the incoming seas, tearing deep clefts through this barrier.

It's in these clefts that the destructive action of the inbound water is evident. About a foot above the surface of the lake, the ice is eroded away by the warmer water and friction until its bulk is cantilevered several feet out over the water; or if isolated as an island, the ice looks like a huge mushroom. The seas roll in and heave up under the projections with a rolling boom to squirt back out. This upward force, along with the tons of weight in the cantilever, finally sheers off sections, toppling them into the lake. These ice "bergs" either bottom out, listing like sailing vessels aground, or float away.

The seasonal succession is not only evident in the layered ice, but also outward into the lake. Older mounds and frontal walls backstop jumbles of thick, sea ice pinned against these former frontal barriers, beyond which a new generation of face ice is mounding up. This progressive ice formation and deterioration along the shoreline is ultimately the product of the variables of temperature and wind.

In periods of warm weather and calm days, the undersides of the projections drip continuously, forming thousands of sliver-like icicles down into the water looking like the mouth of a baleen whale.

While I was walking Muggs on a still and bitter-cold night, I heard the faintest of sounds coming from around the point. I stopped and listened, trying to ascertain its origin. I could have sworn I was hearing the gurgle and tinkle of a brook flowing over and around a rocky course, but there was no such brook, and the creek was in the opposite direction covered with ice. I left Muggs off at the door and hoofed my way out toward the shore. There was no discernable motion anywhere under the sky of a million stars. The lake was totally flat, with the stars reflecting on the surface. It dawned on me that the lake was covered with a film of ice–solid and unbroken like a mirror.

With no perceptible force, a current was moving the entire ice surface parallel with the shoreline, breaking, grinding and crushing the edge of the unbroken ice field against the anchored ice projecting into the lake around the point. In the morning the consequence of this sound was starkly visible. A field of broken glass-like ice, the shards all standing on end, was pushed up in a jumbled mass under the point, actually surmounting low areas of shoreline ice. Jagged sea ice, an inch or so thick and card table in size, was crystal clear, each and every piece tilted every which way but horizontal, and crushed together. Each piece sparkled in the morning sun. Depending upon their set to the sky, they were totally transparent or aqua color. While enjoying this peculiar icescape, I became aware that the ice field was still in motion. The entire flat surface, out beyond the point, was moving at turtle-like speed along the littoral.

## The Paradox of Paradise

\* \* \*

In February we were having three guests up for a weekend of cross-country skiing. By design, we arrived somewhat earlier on Friday night to get things stowed away and organized. The caretaker had turned the heat up from the unoccupied 50-degree setting to take the chill off the interior by the time we would arrive. This time he either had just done it, or something was wrong. The temperature inside confirmed the latter. It was 45 degrees with the heat on. The air coming out of the registers was barely warm. What a way to introduce friends to our new house that I'd been extolling for months. Roger was unreachable, but I left a message on the ubiquitous machine, and stoked up the fireplace. Our introductory bash was held close to the fireplace huddled in coats and blankets, yet nonetheless enjoyable. Our guests brought as a housewarming gift a keen barometer, which I have occasion to refer to each and every day since. I had Jim make a teakwood stand to hold the instrument, and it lives in a very prominent place in the bookcase next to the fireplace. Fortunately, we had lots of extra blankets, but the shower got a rest in the morning. I placed my usual early morning call to Roger relating our night's experience with "his" furnace. I always like to needle him that our furnace problems have occurred because I agreed (reluctantly) to let him install the make he wanted (one I'd never heard of), not the one I specified. Within the hour, three of the four problems were corrected, and the temperature in the Schloss inched upwards. He had to come back the following week to replace the fourth defective part because he didn't have any in stock, belying his past sales pitch, "You ought to let me substitute brand X for Y because I stock the parts, and I'd always have to be ordering Y's parts." I can't tell you how many times corrective work on the furnace has had to await delivery of those "in-stock" parts for Roger's furnace X.

Skiing that weekend was superlative, and, in fact, was so from the first week of December well into March, a season not to be repeated since. Subsequent years brought winters almost devoid of snow, when walking through the forest trails became a necessary substitute for skiing them.

Our ski trails crossed many animal tracks. Those of deer were, in some cases, more like expressways, directional with purpose, used

repeatedly. Smaller animal tracks squiggled across the surface in seemingly random patterns. Often we would hear the winter birds chirping away, and the rat-a-tat-tat of a woodpecker hammering its way through a winter meal. At times the surface of the snow was covered with black dots which looked as though someone had scattered pepper over the snow. A closer look revealed it wasn't dirty snow as the dots were moving occasionally—a bug of some sort. The specific sort of bug, I later learned from someone who knows about buggy things, was a snow flea.

A walk along the snow-covered road on a still night, or a moonlight ski along a forest trail is an experience long remembered. The silence is profound, and the stars are so bright and close in the inky sky. The tunneled vista through the pines and hemlocks is intimate, yet mysterious. The spiritual exhilaration received is beyond the effort required. I never notice my cold toes and prickly cheeks until back inside. Returning to the Schloss from these nocturnal journeys, the creature comforts of the flames in the fireplace and the patina of interior warmth can be seen through the bay windows. Despite wishing the night would never stop, warmth beckons. We just turn off all the lights and let the fire burn out, while sitting on the window seat, staring at the sky until conscious star gazing is overtaken with blissful sleep. This works better than any pill ever invented.

The steep roof is often covered with a deep mantle of snow, which remains as long as the weather stays cold. The well-insulated roof structure efficiently keeps the interior warmth where it belongs. Even with the temperature well below freezing, on sunny days when the south side of the house is sheltered from the wind, it's possible to sit outdoors on the open deck and bake in the sun's intense radiation to the point of being terribly uncomfortable wearing a coat. The bottom shakes on the gutterless roof drip a torrent of water and form long icicles all along the south eaves, appearing like Dr. Zhivago's ice palace. The sun's warmth causes huge chunks of snow frozen to the roof to release their grip on the cedar shakes. They accelerate down the roof, cascade off the eave and smash onto the deck, or thump into the pillows of snow on the bushes. Beware of roof slides on these radiant, winter days! From inside the house, the noise from a rooftop snowslide is disconcerting until the thundering herd of horses clattering down the roof becomes repetitiously familiar.

The bird feeder is a hub of constant activity throughout the winter. It hangs from a cedar branch outside the dining room bay and, from early morning until dusk, is active with chickadees, cardinals, jays and other less-common species, which I have to look up in the bird book. The red squirrels are aerialists that sit on top of the garbage barrel and catapult themselves across several feet of thin air to reach the feeder. Every so often a squirrel misjudges and misses the feeder altogether, to land far beyond. It always brings a grin to see the little miscreant's surprise. But indomitably it hops back upon the barrel to try again. This time though, he studies his trajectory much more carefully before taking the leap. The tops of the barrel and the chopping block are always covered with clumps of cedar cones... leftovers on the squirrel's dining table.

\* \* \* \*

Returning from Muggs' walk the first thing in the morning we turned up the driveway from the access lane. At the same instant, we both noticed a slight movement under the garage's box bay. The drifted snow had created a pocket against the shingled wall, which gave a little shelter from the wind on this cold (10-degree) morning. Fearing that it could be a wild creature sheltering there, I told Muggs to stay put and walked through the trees to investigate.

Under the box bay, huddled shivering on the pine needles behind the drift, was a dog about the size of Muggs. Although visibly frightened, it didn't move when I came close, nor even when Muggs peered down at its nose from the top of the drift–a puppy with black and brown fur and no identification. This led me to think its being there was no accident, so common is the cruel practice of people dumping off unwanted dogs and cats. It responded cautiously to my petting and came along with me into the garage, where I made a bed for it and promised I'd be back with nourishment. Taking it into the house was fraught with all sorts of problems associated with puppies, puddles and rugs. In my defense, the garage was much warmer than the out-of-doors, as it is extremely well-insulated and had been warmed to some degree by my car's engine heat. At least I'd have time to figure this all out. Muggs shared a breakfast of Mighty Dog and Chow with the forlorn little soul. When I set the food dish down, it was emptied in a flash.

Studying the animal, I guessed him (it was a he) to be about ten or twelve weeks old. All puppy fur and huge paddle paws, but a little emaciated, not typically roly-poly. I thought to myself, "you're going to be a big one some day if you make it." And I knew I had to help this placid dog, resting contented in my arms, to make it. Its start in life, up to this point, had not been very positive to say the least, and its future was bleak indeed. Being a Saturday was not going to make things easier, but I began to burn up the phone lines describing the puppy's plight to everyone I could think of. Joyce said I should call the radio station. They volunteered to put an announcement about the lost dog over the airwaves. Debbie, from PJ, said she'd make some calls to try and find it a home, as did Barbara, my caretaker. I called the sheriff who wasn't interested in my problem, but gave me the number of the pound. Taking the little creature to the pound would probably be his death sentence. To the credit of the keeper, I found out later that he is a very sympathetic man who does his best to find homes for the animals incarcerated there. I had visions of this beautiful dog (either German shepherd or a blend of shepherd and something else–it's hard to tell when they are so young) being sold to a "buncher"; a dealer gathering animals for resale to labs–an end no living thing deserves.

I also called neighbors living along the Drive year-round to see if they recognized the dog or knew of anyone that had a new puppy such as this. I somehow knew there wasn't going to be any call from a distraught owner. It's one of those sinking feelings again. Phone call after phone call with no luck. Debbie called back with the same results. She thought I ought to call a boarding kennel in a nearby village. Anything's worth a try—you've got to "network" as the young professionals call it. The owner said that I ought to try the humane society in Green Bay. If I could take him to Green Bay I could take him home. Anyway they were closing for the weekend.

The kennel gal said I should call her back if all else failed. When I did, she said she was just about to call me. My hopes rose and stayed that way when she said she was having a friend call who would give him a foster home until she could find someone suitable who would like him.

During the day I took him dog biscuits, breaking them up into several pieces so his puppy teeth would have an easier time. He was curled up sleeping when we entered, but perked up quickly and got the hang of the biscuits, which were evidently foreign to him. Muggs stood by his bedside, tail wagging back and forth like a metronome, watching the goings-on. She realized that this time the biscuits weren't for her. The three of us went outside for some exercise until the puppy's antics got to Muggs, and she let him know to lay off in no uncertain terms.

I'd given the foster care volunteer directions to the Schloss. She'd replied she would "pile the dogs into the car and be down by 4:15." We were outside playing when the car pulled in. I looked down the driveway to see a beat-up old Scout whose windshield was filled side to side with dog faces of every description—and one small young-looking woman squashed in the driver's seat. Dogs and woman all looked at us approaching the car. I wasn't aware there were more dogs in the back seat platform, because the windows were so slobbered up.

Rory gave a friendly wave and energetically got out to see her new charge. She was dressed in a sweatshirt and pants tucked into rubber boots, long, stringy red hair flying every which way. She related that she owned a farm by Heins Creek north of Jacksonport and spent her time caring for injured animals both domestic and wild. I asked her how many animals she had.

"Twenty-three dogs (not all twenty-three were in the Scout, it only seemed like there were), eighteen cats, three horses, two llamas, a donkey, three goats, two deer and two seagulls."

All the dogs and cats lived in the house, the second floor devoted to the cats and the remainder for her and the dogs. She was an accredited wildlife rehabilitation person. The wild animals under her wing had been injured and were being rehabilitated for reintroduction to their proper world. If ever one could sense that a person had a way with animals, Rory had it. She opened up the rear of the Scout to let the dogs get acquainted with the puppy. Without any barking, the dogs (some were huge and all were clean and well-groomed) each quietly and gently nuzzled the little puppy cradled in Rory's arms. A couple of them gave it a big slathering lick on its face. All the time Rory softly discussed everything with the dogs, who were attentive to her every word.

By looking at the puppy's teeth, she confirmed that he was probably around twelve weeks old, and observed that he hadn't had much to eat for a while. His coat of fur was good and his eyes clear, but very tired. She said she would deworm him that evening and get his puppy inoculations on Monday.

Rory had been instrumental in establishing a county-wide humane society. As they were looking for land (and money) to build a building, I made a donation to the society and also one to Rory to help buy thirty-six pounds of daily dog food.

When she drove away, the puppy was sitting in her lap, and the other animals were settled in for the drive back. I felt a confusion of emotions. Heartfelt gladness that this beautiful little animal we discovered now had a good chance at life, rather than one or another of many premature deaths. A tinge of lingering sadness from the experience of finding an apparently cast-out animal, lost and bewildered, hungry and scared. Anger at the thoughtlessness of the person who had won the trust of the animal only to cruelly abandon it (for whatever reason), leaving the quite helpless creature to wonder what happened that those it trusted betrayed the trust—as only a human being can do. I felt some sadness that my life is such that it wasn't realistic I could take in this animal and make it a part of our home, and experienced relief that I didn't have to deal with how I was going to manage to take care of him. Because in the end, I don't think I had the heart to take him to the pound. And lastly, thankfulness there are people in the world such as Rory with hearts as big as a house (or a farm in this case) and a selfless devotion to caring for other creatures, less fortunate and innocent of life's predicament, whether they be animals or children. Both need lots of love and care, and those who give of themselves this unmeasured love are each in their own manner, saints of sorts.

I remember relating this incident to a friend who exclaimed, "She must be nuts!" Reflecting on that perception a little, I seemed to recollect that throughout history common wisdom often has confused uncommonly unselfish devotion with aberrant behavior and vilified or martyrized those who haven't fit the mold. Maybe Rory doesn't fit the ordinary mold, but God bless her for her unusual character and her determination. When I think of that dog's sad little face, it prompts within a conscious renewal of my feelings for my charge, and I give Muggs an extra hug to make sure she knows how much I care about her.

No matter how beautiful the winter is, nor how interesting the activities are, by late February the effort of bundling up in heavy coats, boots, woolen caps, long underwear, and the like to enjoy the out-of-doors begins to wear thin. Contrarily, a heavy snowfall at this point of winter doesn't have the ominous implications it would in December or January. The texture of the clouds on sunny days in March seems to be different, and the blue of the lake has subtly changed its hue.

The lengthening days of late February and March seem to be a signal to the birds, changing their vocalizations from necessary communication to a melodious celebration of nature's new year. The cardinals must feel a similar exhilaration as mine when I am feeling the sun's returning warmth through my jacket. Their joy is simply more demonstrative.

## Chapter 14
# Spring

It's wonderful to walk along the untracked beach feeling the sun burning through the ski jacket into my back, tantalizing me with the promise of better things to come. The warmth of springtime that first year took a long while to clear the forest floor of snow, yet the sun's power on the south-facing beach rapidly exposed the sand. The swamp behind the house filled with melt water several feet deep. The last vestige of the ice armor was pockmarked with holes in honeycomb fashion where the ice under the stones melts much quicker. The face of the ice had receded far above the ordinary waterline. Huge sections of this senescent ice had cracked off and were tilted toward the lake. Rivulets of melt continuously wriggled down the sand into the lake. The creek was flowing with renewed vigor, draining forest and swamp along its course. It was clear of ice and snow except on the high, shady, north-facing banks, yet the ice bridge over the stream at the shoreline remained. Once the ice bridge collapsed, it would effectively restrict our walks along the beach until the stream's level dropped, and its current subsided; when summer weather would permit wading through the water with bare feet.

Early that first spring in a gap in the remnant shore ice, I noticed an animal washing to and fro just off the beach. I had my sea boots on so I waded in to investigate. It was a large dog with a collar and ID, "Maggie Pie," an English setter. A phone call to the number on the tag netted the information that Maggie Pie had been someone's pet in Sturgeon Bay, being cared for by her brother who lived on the Drive. The dog had disappeared over two months previously. The dead dog was stark proof of how dangerous the shore ice is. Slip off the top of the ten-foot-high wall into deep water, and that's "all she wrote" for man or animal, drowning in the icy water or smashed senseless against the wall, time and again, by the incoming seas. Or simply fall through a crevice—or punch through weakened ice—other ways to be done in. I am so watchful of Muggs, never letting her out on the shoreline ice.

Nor for that matter on the ice over the creek, which is often far above a receded water level. In an instant, its current would have her under the ice downstream of the hole and me helpless to reach her in the tunnel. Anyone living in the vicinity of the shoreline who lets their dog roam unsupervised in winter does so out of ignorance of the danger, or disregard for the pet.

My fear of Muggs falling in the creek was not groundless. One early spring day, the only ice remaining in the creek was a snow-covered ribbon along each bank. Muggs had been poking along behind me on the path beside the creek, but when I turned to look for her the path was empty. Out of the corner of my eye, I noticed a couple of concentric ripples spreading upstream on the calm, black surface of the water. It took but an instant to realize those ripples could only have been caused by something dropping into the water. Tearing back along the path, I found the frightened Scottie standing on her hind legs, up to her neck in water, futilely pawing the vertical bank with her forepaws. Where the path had dipped close to the creek, she had stepped out on the ice. It broke, and she plunged in head first, her short legs woefully incapable of pulling her up on the bank. They couldn't even reach the top of the bank. Her eyes said it all as she caught sight of me reaching down to pull her up by the collar, sufficient to reach under her for the big boost. She was a forlorn and subdued creature in the laundry tub getting the mud washed off and fur dried.

The beach always needs a cleanup of winter's accumulated detritus. It is incredible in variety and disgusting in disregard for the environment: styrofoam cups, shoes, pop tops, plastic of every description, tangles of monofilament fishing line, fishing lures with ugly killer treble hooks, shotgun shells, torn sandbags, beer and pop cans, hats, gloves, straws, six-pack wraps, jars and other unmentionable items of jetsam washed ashore. It all goes in a garbage bag so that we can enjoy a waste-free beach—at least temporarily.

Impossible to clean up, ugly to look at and noxious to smell are the occasional streaks of glistening black, oil-soaked sand. This pollution often occurs at times of the year not associated with pleasure craft. My guess would be the oil is from commercial vessels' bilges being discharged overboard. I'm probably naive, but it's hard to imagine it being purposefully pumped out by U.S. or Canadian vessels regularly plying the lakes. These people generally seem to have as great a concern and caring attitude for these inland seas as do any of us, whose consciousness has in the past generation been awakened to a different environmental ethic.

I have only to look at the change in attitude that has occurred within me in a generation or so, to have an insight into the broader awakening of most of the entire community dependent upon these waters. As a teenager sailing in the '50s, everyone dumped their garbage overboard with not the least concern. Admittedly there was little plastic used at that time, but it wouldn't have mattered anyway—the consciousness just wasn't there. We were simply doing what our predecessors had always done... "It all sinks eventually." Think what the bottom must look like! Yet when racing smaller boats, we always kept a cup onboard so we could dip in the lake for a drink on a hot day.

By the '70s, the number of vessels using these waters had increased exponentially. The opening of the St. Lawrence Seaway brought in hundreds of large ships. Dozens of foreign vessels crowded these ports and could be sighted on these waters on any given day, either anchored offshore awaiting berths, or underway. The variety of foreign flags was interesting and exciting, but there was a price, and the lakes were paying. The advent of fiberglass brought yacht ownership within reach of other than the very wealthy. Manufacturers were popping boats out of the molds as fast as they could to satisfy ever-increasing demand. Existing yacht harbors filled quickly to beyond capacity. For a new owner, getting a mooring involved lengthy waiting lists, or who you knew, or who you could pay off in many cases. Industry along the lakes boomed and flushed its poisons wantonly.

Those of us sailing upon these waters began to notice change on our beloved lake we so took for granted: large oil slicks, human excrement, chemically discolored water, noxious-smelling harbors, dead and dying fish and birds, malformed fish and birds, less fish and birds. Rookeries I had walked along, on remote beaches and islands that had supported thousands of birds in the '50s, twenty years later had only remnant populations, or were gone. The water quality along Door County's Green Bay shore was visibly tainted. It didn't look good until further north. Rachel Carson wrote *Silent Spring* and said it all in stark terms. Scientists from our universities spoke out. Periodicals hammered away with bad news. And finally, most of us woke up to our horribly cavalier treatment of these waters we loved. "We got religion," and you know how fervent converts are. Thank God, because battles against new environmental enemies of the lakes are always surfacing.

The water quality of our lake is visibly better, and the wildlife supported by these waters is increasing both in number and variety, but the most insidious chemical pollution remains vexing in eliminating the sources and cleaning up. When I walk along the beach and see the stretches of oily-sand derivative of a single pumped bilge, I shudder to think of the environmental catastrophe resulting from a tanker collision or grounding.

\* \* \* \*

I awoke, on my birthday in late March, to a sound long absent—the cry of gulls wheeling back and forth over the shoreline. It was a sound from a source I could not see as the atmosphere was thick with fog over the lake and veiled the tops of the tall pines. The stillness on the beach was absolute, except for the occasional squawk from the gulls. As far as I could see down the beach into the gloom, all that remained of the ice armor was a Gibraltar-like promontory off the point, and faintly, an ice island offshore with two gulls perched on top.

To be sure, the ice armor was still underfoot but covered with sand and pebbles. A humpy, bumpy ribbon running along the shoreline, with a texture foreign to the smoothness of a sandy beach, which normally seeks its own level as it is washed over and over again. No mistaking this for what it was. Underfoot it was rock solid, not squishy like treading on sand. The surface of the lake was mirror smooth. The bottom clearly visible; stony just off the water's edge and falling into deeper water with row upon row of sandy ripples, outward and downward until lost to sight by reflection. Off the point, the rock bottom reminded me of a Roman road of antiquity, paved in patchwork fashion. The earth's crust was scrubbed clean, exposing an underwater patina of raw-umber rock shelving off into the deep.

The fog touched all, glistening on the twigs with their buds, on the stems of last year's beach grass and on my wool hat. It felt wet and cold on my cheeks.

Out on the lake a mechanical sound became just barely audible—the metered throb of a boat's engine in slow motion. It passed gradually over the sound curve and faded away in the opposite direction. I, so secure ashore, was enjoying the beauty of the scene, yet the mariner in his pilot house was feeling only anxiety, cursing his blindness caused by the invidious fog.

It remained a cold and clammy day; our world constrained by a very limited range of visibility until afternoon brought with it a hint of blue directly overhead, a harbinger of change. As the evanescent pea soup thinned and eventually burned away, the coastline emerged in a blur, slowly brought into focus until the shoreline separated from the water's surface as it reached the extremes of vision. The temperature soared with the clearing sky, bringing warmth unfelt for months, born of the sun's post-equinoctial power.

My thoughts were that I was blessed to be able to be where I was, doing what I was doing on this windless "see-forever" day, rare in the month of March. The afternoon's warmth not only did its trick through my jacket, but also on "Gibraltar." Hour by hour it was slowly sinking until its sloping back side was awash just off the shore. With the night staying above freezing, by the next morning this large ice promontory was a mere fraction of its former bulk. By the time Muggs and I reached the beach for our walk, all that remained of this boxcar-sized monolith was a zillion ice cubes sparkling in the water, and one table-top chunk beginning to float away.

* * * *

The frogs of springtime have no peers as noisemakers. That first spring's extra heavy snowmelt filled the swamp to the brim. The brim being the level of the access lane at one end and the shoreline ridge opposite. Most of Tracts 15 and 16 were under water. The crown of the shoreline ridge was the only area high enough to be dry, and this was only 30 or 40 feet inland from the beach. Our bedroom looks out over this seasonally flooded scene. Though it was wonderful to have the window open at night to enjoy the redolence of spring, for about two weeks we thought we were living at the edge of the Okefenokee. The croakers, from bass to soprano, were having nightlong song fests. Dozens of them drowning out even the surf. No measure to this music; a riotous cacophony building to crescendo—then abruptly silent. Only to start over with the choirmaster tuning up.

The following two years were ones of severe drought that dried up the swamps and the creek. Never since have the frogs of spring been in abundance—only a duet or two. I never thought I'd miss that nightly torture, but I do. I wish they were there to force me to shut the window again.

When the frogs' chorus quit later in the season, the silence of the nighttime swamp was a welcome interlude. It turned out to be only an interlude because shortly thereafter, a pair of ravens built a nest high in a topped-off pine just outside the bedroom window. The ravens arose with dawn and began loud, raucous conversations lasting until my breakfast time. As first light was arriving ever earlier, my sleep was becoming ever more abbreviated. These ravens must have been morning creatures... I'm not.

On sunny days I gave up trying to sleep and tagged along with Jan and Muggs on early morning walks on the beach. Neither one moved very fast. Jan was looking for knotholes that had washed up on the sand. Over the years she has collected these pieces of wood with holes in the middle, the remains of cedar tree trunks or branch stems, where once joined to a branch or twig. Of every size and shape from a dime to a small saucer, some are almost all hole with little surrounding wood. All are polished smooth from washing back and forth in the water along the beach. The collection has grown to fill three wicker baskets.

Toward the other end of the property, the creek has its own springtime calendar of events. Fishermen are constantly stopping on the Drive to look longingly at the creek. The large culvert, which carries the stream under the road, has always been a mecca for people who slide down the shoulder and stand upon its projected end. It's a great place to see if there are any fish milling about, or just to look out over the floodplain scene downstream as the creek meanders toward the beach.

The water level of the creek as it nears Lake Michigan is predominately a function of the level of the lake. During high-water years the damming effect of the surf backed up the creek well over its floodplain at times. In fact, one spring the creek rose to within six inches of the Drive's pavement at the culvert. The creek level receded when the lake did, and exposed its floodplain to view. It also began to regenerate growth of a profusion of grasses and wildflowers on this low-lying land.

Not wanting fishermen to trample on the emerging growth and thinking many folks might not realize this was private property, I placed "No Trespassing" signs along the road with mixed success. Those people who normally respect the property owner's wish don't trespass, while those with inalienable prior rights do what they want anyway. The fishermen are looking for the annual run of smelt, a small herring-like fish that run in schools of thousands up the creeks from Lake Michigan. If you are in the right place at the right time (for a week or so) you can fill tubs with them. Smelting always seems to be done at night and is as much a social event as a serious fishing endeavor. For those of us who live on the beach at a creek, spring brings unseen visitors wielding flashlights, nets, buckets and beers. If the bobbing lights and voices on the beach stick around, it's good for them and bad for us and the smelt. In the morning the only evidence of the nighttime smelting is the myriad of footprints in the sand at the creek and sometimes empty beer cans.

Other fishermen scan the creek in vain for runs of trout, but all they find are suckers, which from above look like trout to those of us who aren't anglers. For a period of a week or two these fish come up the creek to spawn by the hundreds. My first exposure to this spawning run was at night walking Muggins along the beach. When we got close to the creek it erupted. In the darkness, I thought deer were splashing across. Shining the flashlight on the creek, it illuminated a solid mass of fish slowly finning in the shallow stream, so shallow in places that the fish were only partially immersed. Our nearness had caused a mass flight upstream. I, of course, thought they were trout. Trout or not, this instinctive springtime swim against the current, is fascinating to watch on a sunny day when the water is penetrated by light. Scraping along the bottom, they fin their way up through the delta-like shallows at the shoreline. A sprint, a rest, a sprint, a rest and a final dash to deeper water. They hide under logs or shady banks and work their way through the riffles upstream, some as far as the deep pool, where the tea-colored water rushes out of the culvert. Try as they do, they can't seem to make the leap up the water pouring out the culvert. Strange thing is, I don't remember ever seeing any go back downstream after their egg-laying stint in the sandy bottom.

By the end of May they've all disappeared, other than one or two dead ones floating upside down. This posture exposes to view the ugly platelike lower jaw with lips around the rim. That's how this neophyte learned they weren't trout. Until this letdown, I was so taken with my trout stream that I didn't tell a soul for fear I'd have more fisherman in the stream than fish. As it turned out, everyone around these parts, but me, was aware of the phenomenon occurring and the nature of the fish.

The first really warm day of that spring brought Rick and his bulldozer back to do the final grading above and below the seawall. By the end of the day the machine was gone, and the rakers had finished it off. I was very relieved to see the Cat lumber out the driveway and onto its trailer, to be hauled away forever. The finished product looked rather smooth and urban, but not for long. The lake was significantly higher than the previous year. Storms soon did their work recontouring the beach, pushing sand against the sheet piling, and then over the top. Within several months, there was no wall visible for most of its length, and it has remained so ever since.

Above the seawall and up to the junipers fronting the house, the sandy meadow was barren of vegetation from all the construction activities of the previous year. Within weeks, shoots of long-stem grasses began to show up to begin the regeneration of the meadow. The notable exception was where Don had burned the tornado debris. This area would take years to grow anything.

To provide some added character to the sandy meadow, I had Greg plant two irregular groupings of five Bar Harbor junipers. The junipers are an attractive complement to the tan meadow grasses. Over each summer I have been very energetic in my weeding. Pulling up all broadleaf weeds amidst the grasses, and the individual stems of grass growing up through the junipers, until the character of the meadow is much as I first saw it—a thick variety of tall grasses; now with the added touch of the junipers. These low evergreens have long since merged together and spread wide across the sandy soil. We added a third grouping of thirteen, strategically located to brake blowing sand driven diagonally across the meadow by southwesterly winds.

# Chapter 15
# High Water

Twelve months after the house was completed and only eight since the shoreline protection work was done, I again began to be preoccupied with the higher-than-ever lake level. Not that the seawall wasn't doing its job, but there were some potential problems beyond where it terminated at the point at one end, and associated with the creek and slough at the other end. To the northwest beyond the point, the only thing keeping Lake Michigan from inundating George's property was the shoreline ridge. Should the lake in its awesome power breach the ridge, I would find Lake Michigan all the way inland to the access lane.

Exploring through Tracts 15 and 16, I made the startling discovery that the shoreline ridge had been breached not by Lake Michigan, but purposefully by man. It became apparent that my nearest neighbor (or a previous neighbor) had long ago excavated a gap in the ridge to drain his otherwise undrainable property. This made some sense as the previous spring's flooding covered much of the neighbor's property with up to a foot of water surrounding an unoccupiable house. However, there were several problems with this drainage solution. Foremost was the use of another's (George's) property to solve a poorly engineered site development. The nuance of this illegality apparently had never occurred to the trencher.

If ever the gradient of the trenching ran toward the lake, it wasn't by the time I began to observe it. The lake over time had deposited sand and stones in the gap, up to 15 or 20 feet inland. The "drain" was overgrown, as were the trenches. I stood at the edge of the gap several times on days when the onshore breeze was blowing a moderate 18 to 20 miles per hour. The small seas were washing through the gap and down the trench, actually filling the swamp with lake water. It didn't take an expert to figure out that in a gale the breaking seas would begin to chew away at the sides of the gap, increasing the breach in the dike, so to speak.

To illustrate the phenominally high-lake level, that fall we decommissioned *Aurora* which was moored to a Palmer Johnson pier four inches under water. We sloshed back and forth across a boatyard surfaced with the bay even into some of the buildings. This was getting serious! And all the experts were predicting further increases.

My concern was not limited to this problem on George's property. Towards the southwest end of the sheet piling, seas were at times washing over the top of the land behind the sheet piling and draining down into the slough. The slough had become a large pond and was no longer separated from the creek. If Lake Michigan did continue to rise, my concern was that the overtopping seas could ultimately wash out the sandy soil behind the sheet piling and into the creek.

I decided I'd better prepare myself for the worst. At my request the DNR representative came up from Green Bay to take a look at things. It was important to determine the permissibility of the improvement, should I decide to proceed with additional sheet piling. My concept was to continue the run of sheet piling from where it hooked back at the creek. The intended continuation would run perpendicular with the shoreline along the nearside of the creek and die into much higher ground in the forest. The slough itself would be filled to create a new line of dunes between the beach and the front side of the higher, forested land behind.

The most significant comment made by the DNR person was not about my sheet piling proposal, but concerning the breach in the shoreline ridge. He saw exactly the condition of my concern because the lake was surging through the gap as we stood there. His words were, "This is a potentially dangerous situation, and something should be done to correct it before it's too late."

I called George, explaining the situation. As the reader might imagine, George wasn't about to spend a nickel to do anything on a piece of swampy property that was a white elephant. Not that he described it in any such terms. I proffered that I was possibly going to do some other work and would he mind if I had the gap closed up by my contractor, while he was on the job.

In that it was infinitely more important to me than him, George probably had presupposed that I would get around to this, offering to do just as I did. "Oh well!" He readily agreed to this proposal, as the value of his swampy property would be zip if Lake Michigan destroyed the shoreline ridge.

Down deep inside I knew what had to be done at the southwest end of the sheet piling and I began the mental gymnastics to prepare myself for the onslaught of equipment, tree protection, etc. Deja vu.

Rick's initial activity was to fill the slough to a level just above the creek to provide a working platform. For three days a large front-end loader scooped sand and gravel out of the lake at the waterline. For three days it scooped from the same spot with the lake filling it in as fast as it was removed.

While this was underway, the bulldozer's scoop was loaded with rocks the size of volleyballs. It churned its way in the lake around the point to fill the breach in the shoreline ridge on George's property, to my immense relief.

There I was only a year after the house was completed and into another major project. By this time I may have been getting inured to the disruption. Thankfully, the people doing the work were onboard with the program. But, with certainty, I can attest that I was not inured to the costs of all these engineering endeavors.

The reaction from folks up and down the Drive was as interesting and diverse as human nature itself. People whose properties were suffering erosion or exposed to the potential of this condition came to look and ask, as they agonized over their dilemmas. Spend the money or wait and see? Those who lived on rocky parts of this stretch of coast were smug in their erstwhile wisdom of building upon or owning rock-bound coastal property. Others watched their shoreline being eaten away and opined that I was a shade loony to be spending all this money. One neighbor to the southwest commented that he felt my seawall had been the singular thing that prevented his beachfront from eroding.

On the other hand, our immediate neighbor's wife seemed to think the seawall was the direct cause of the creek bending across their beachfront before emptying into the lake. It certainly must be an aggravation to have to wade through the creek with its high bank to reach your summer fun on the beach. Despite this, I don't understand her reasoning as at times the creek flows in the opposite direction, hairpinning as much as 200 feet across my beachfront. They spend the summer trying to redirect the creek and fill the dry bed with beach sand, shovel by shovel. But the creek has a will of its own, driven by the wind and seas off Lake Michigan. It may well change direction a half dozen times before we see these neighbors the next Memorial Day.

The actual work progressed rapidly as there was no trenching required. The finished top of the sheet piling was three feet above the working platform of fill. All the welding and the construction of the concrete anchorwall were done on the surface. After everything was removed from the site, Rick brought in forty, large truckloads of sand, which was graded up over everything to create the dune. At noon on that day the first snow began to fall and, several hours later was several inches deep. The rough grading was finishing just in time. Again, it was left over winter, and the next spring they came back with an additional fifteen truckloads of sand to create the final shape of the "new dune," as it had come to be known. Across the top, Greg transplanted several large cedars, two spruce and a variety of aspens and birch. Along portions of its sloping sides clumps of Bar Harbor juniper were planted to create a wind foil. The dune was seeded with tall grass and given a jump-start with some transplanted clumps of grass.

I had had some misgivings about our ability to create this natural feature and landscape its crown artfully enough to look like it had been there forever. After it was completed, I told Greg his skillful replication of a dunescape ought to have a prize. Seven years later the coniferous trees have grown together, the grasses and raspberry bushes are a yard high, and a variety of wildflowers bloom in profusion the entire growing season around the ever-expanding carpet of junipers. And, I've never once missed the slough. My immediate feeling upon completion was that I was done, once and for all time, with these bastions of shoreline protection. So confident of this was I that I had Greg plant a row of large seagreen junipers along the extension of the driveway down to the beach. My purpose was to reduce this "truck route" to the width of a path and to hold back the onslaught of thimbleberry bushes, which were proliferating along this area. Never again would the driveway be used for heavy equipment.

\* \* \* \*

The barometer on the bookshelf gives, at a glance in passing, a subliminal awareness of weather. It's much better than the local radio station, which lets its listeners know when it's raining. My cognizance of the barometer may well stem from its use in a lifetime of sailing, but I must confess that reference to the instrument at sea in recent times has been somewhat supplanted by more frequent and increasingly accurate marine-weather broadcasts. It's only when *Aurora* is out of radio range that I turn to the barometer. At the Schloss I've come to rely on its unerring forecast.

On a Saturday in February in the high-water year of '87, my passing glance registered a significant drop from morning to afternoon. By evening its drop had become a scary nosedive, and it had not stopped its plunge. I commented to Jan that I was concerned with what might be heading our direction in the way of wind. At bedtime the barometer read lower than I had ever seen. At 2 a.m. the storm struck like a sonic boom against the side of the house, awakening me out of a sound sleep. Daylight gave vision to the shrieking winds and a serrated horizon foreshortened by mountainous seas in perpetual motion. With our south-facing shore, the north wind posed no danger from storm surge, but I wondered what this 65-mile-per-hour gale (abetted by an abnormally high-lake level) was doing to the shoreline at the south end of the lake. WGN, a Chicago radio station audible in Door County, told the story of flooding over Chicago's lakefront parks and nearby streets–conditions never before experienced; of severe damage to lakefront buildings and Lake Michigan surging far inland from its normal shoreline. The impact on Chicago was televised nationwide. Fortunately for the Schloss, the worst storms on the lake are typically from the north, but the damage along Chicago's lakefront, seen so starkly on TV, did little for my peace of mind. In the extreme high-water conditions even the change in level caused by low barometric pressure can cause flooding.

On the other side of the coin, the extremely dry winter throughout the lake's basin didn't square with the experts' projections of even higher water the following year. Not being able to measure any sort of water level on a beachfront, every time I was at PJ's I would look at the water level pier-side, where they were agitating the bay to keep it clear of ice around the pilings. As winter lapsed into spring, it was not rising. If anything was occurring, it was dropping—an unusual, if not unique, circumstance in this season of snowmelt and rain.

When we had been looking for lakeshore property along the Drive, availability of lots and homes for sale could have been counted on the fingers of one hand. The high-water conditions seemed to open the floodgates of available shorefront real estate. I counted thirty-three properties for sale along the seven-mile Drive that winter. And they weren't selling. This year-or-two aberration has been the only pause in an otherwise steep upward curve in property values. Recently, a neighbor lady driving by stopped to chat and was gloating over the most current and exorbitant sale price of a vacant piece of shorefront property nearby. She and her husband are retired full-time residents, so I asked if they were planning on selling. When her answer was negative, I replied that she must enjoy paying the astronomical increases in property taxes. For those of us intending to stay around forever, inflated values simply mean increasingly burdensome taxes. The good ole boys running the local government and school districts must love it. The shoreline property owners pay the bulk of the taxes in the town, and for the most part do so without any representation. There are only thirty registered voters along the Drive; the vast majority of us being second-home owners.

# Chapter 16
# Planning Ahead

I had, by placement of the rock in the breach of the shoreline ridge, stopped any immediate threat to our property from that direction. Still, I experienced a nagging concern over the longer term implications of not owning Tracts 15 and 16. Residential development in the county was escalating, and more marginal lands were being built upon. The County Zoning Administration was not one of dynamism, being largely reactive. And a reactive regulatory posture is simply one of closing the proverbial barn door after the horses are gone, as so many communities have belatedly discovered.

Because the high dunes along the Drive are a unique feature along the west shore of Lake Michigan, I suggested an ordinance to the department that would prevent building in front of the tree line. This admittedly restrictive (in some cases) ordinance would have insured the preservation of the preeminent nature feature of this coastline, the very thing which gives it its unspoiled quality. Up to this point of time, no one had elected to build out on the open dunescape, despite that not doing so had placed some of the houses on the rear one-tenth of the properties.

The County Zoning Administrator did draft an ordinance and present it to the appropriate legislative committee. Although this type of restriction is relatively common in other areas, the proposed ordinance never got out of committee. They felt it couldn't be applied throughout the county. It wasn't intended to, being focused on the preservation of this unique natural feature, only nine or ten miles in length. Within a year someone violated the tree line, placing a house out upon the open dune. The uninterrupted tree line was broken, and adjoining neighbors now had their vista of the coastline blocked by the bulk of their new neighbor's house. What followed, of course, is that the next home owner-to-be wanted to locate his house even further out on the beach, but needed a variation, which fortunately wasn't granted. If this sequence were allowed to continue until the last property was developed, the evident consequence is that the natural feature, which gave the place its original beauty, would be obliterated behind a wall of architecture as diverse as the tastes and pocketbooks of the home owners. Everyone has seen pictures of shoreside communities such as this—not an enticing advertisement of sound land use.

While on the soapbox, I will inveigh against another expedient and destructive development practice which I became aware of during my search for property. It should be required that properties be developed within the context of the existing natural contours; rather than simply slicing off the dunes' crown to fill the swales, which results in a virtually treeless tableland. This practice alters the entire landscape and leaves large, gap-toothed interruptions on the forested beachfront. This is to say nothing of the ugliness of the filled swales from the perspective of adjacent properties, nor the total destruction of natural drainage patterns.

It is voiced by the cognoscente that the county government will do nothing that it considers inhibiting to the construction industry. Building is, ironically, second only to tourism in creating earned income in the county.

I had visions of George finding his "pigeon." The pigeon then to do nightmarish things to make the swampy 15 and 16 developable. I undertook a topographic survey of this land to illustrate with the contours the extent of seasonal flooding. I sent this to George, not with an offer to purchase, but ostensibly to illustrate the seriousness of the breach problem. My underlying purpose was to get him to think about the whole matter. Predictably, it elicited no reaction. George just wasn't ready to deal.

The second winter was as dry as the first was wet. The land was barren of snow cover. Flying in the face of all the experts' computer modeling, if not historical precedent, Lake Michigan's water level did not rise in the spring, a most welcome relief to the anxieties felt by all who had waterfront interests. In fact, that summer and fall the rate of drop in water level was record-breaking, aided by severe drought conditions throughout the lake's basin. The width of the beach in front of the Schloss expanded by at least 50 feet. Off the point, the receding water exposed extensive formations of bedrock quite far out into the lake off the point. For the first time, we could actually walk around the point without getting our feet wet.

Over the next couple of years as the lake level continued to fall, also did the creek, scouring out its bottom and deepening its banks. Just 20 or 30 feet in from the beach, it exposed the timber foundations of a bridge which had been a feature of a long ago "Drive" when it was nothing but a track in the sand.

It was interesting to watch the evolutionary regeneration of the beach. Where the water had receded, the shelf rock jutting into the lake began to be covered with sand. The littoral drift deposited incredible amounts of sand on either side of the point. The incoming seas washed the sand higher and higher on the beach, and their oblique direction spread the sand farther along the beach away from the point. Higher up on the beach, wind-driven sand filled low spots and created windrows out of which grasses began to emerge. And then a sprout of sea grass showed up–to multiply, and spread and thicken along the drop-off just above the high-water mark. The clumps of sea grass gripped the blowing sand and began to transform the leveled-off beach into random berms, covered with grass three feet high. Other pioneer plants, such as mullions and alder, began to appear and thrive in the inhospitable sand and pebbles of the beach.

It took two years, but George finally did respond. With the lower, safer water level I thought his incentive would have been lost. But each of us has our own triggers. George's was advanced age and a sick wife. He wanted to be rid of this potential problem. We reached an agreeable price on the phone and expected to conclude the deal within a month. But the deal hit a snag, unhappily for George, because it cost him some money in attorney's fees to clear up.

This purchase, which looked so straightforward, was to take six months. I had a land survey done for the property transfer and the surveyors discovered that my nearest neighbor's garage was eleven feet onto George's property. The neighbor's mound-type septic field also intruded. As I understood it, that property had been acquired by the current owner in a mortgage foreclosure, and she was trying to get rid of it. Although I can understand her reluctance to dump in more money, it certainly wasn't going to sell with an encroachment to cloud any title commitment. Nor was I going to buy any property knowing the problem. It took George and the neighbor several months to iron this out, with the inevitable result being the forced purchase of an 11-foot strip to cover the encroachments.

Maybe it's difficult to understand anyone's unmitigated joy in having bought a swamp, but I knew that feeling. Now, when I looked out the bedroom window toward the northeast, as far as I could see through the trees was mine. No one could intrude upon the savored isolation, nor fill the swamp. The sylvan vista from the bedroom window was secure.

The path leading through the arbored forest to the point is one of my favorite scenes to enjoy in every season looking out the bedroom window. It is just as appealing walking along the path toward the house—sort of like looking through the wrong end of a telescope. The focus at the other end is on a corner of the building sitting at the end of the path, solid and comfortable in its place in the woods. This can be ever so inviting in the chill of dusk, with a dusting of snow on the ground, smoke curling lazily upwards from the chimney and a soft amber light glowing through the bedroom window.

If it seems like some la la land described by an unregenerate romantic, I suppose it is, but I am, and the scene never ceases to cast its spell. I like serenity and beauty in contrast to conflict. There is more than enough conflict and acrimonious debate going on in our everyday business world. I have no problem dealing with this sort of thing as a necessity, but it's not my thing. It reminds me of my rather lengthy relationship with a client, whom I saw regularly for several years. The president of this Fortune 500 company delighted in pounding his desk and letting us (and others) know in stentorian tone that, "I don't get ulcers—I give them!" Romantic, idyllic images are good for the spirit, so why not our designs, be they buildings, or dreams, or goals.

It was fall before the ownership of Tracts 15 and 16 had transferred to us. I'd time to think about what I wanted to do to make them better. Large trees that had toppled during the tornado needed to be removed—two of them across the path and several out onto the beach. There were some standing, dead birch which were candidates for cutting, as in falling on their own they would be dangerous to other trees. I like to leave the trees standing for birds, but paper birch rot so quickly that within a couple of years they menace other healthy trees, which the birch might damage by their uncontrolled falling during storms.

By far, the greater problem to deal with was the need for shoreline protection, though the gap in the shoreline ridge had been closed up with rock. This had done the job from the standpoint of keeping Lake Michigan where it belonged, but it left an ugly scar. I didn't like the overgrown trenching slicing through the woods either.

In a predisposed vein, I began looking at the eroded tree line embankment. My stone seawall ended at the point at the former property line. The newly acquired beachfront was unprotected, and the end of the existing seawall was vulnerable from northeast storms. The water level was down over two and one-half feet, and more beach was being created all the time. Since there was absolutely no danger (nor might there ever be any), the more practical approach was to leave everything status quo; waiting out any high lake levels in the future.

For me it wasn't so easy. Should the work be postponed until higher water was again chewing into the shoreline, the contractor would have no place from which to operate his machinery. They'd have to cut a strip of trees fronting on the beach across the width of the property. This would totally destroy some of the best trees on 15 and 16. As it was, if I could find a route to get the machinery and rock out onto the beach, the contractor would have the wide beach to use as a working surface. This was the real dilemma, how to get from the access lane to the beach without creating a road through the forest and swamp on 15 and 16. There was no way other than using the driveway to get out onto the beach. But, once on the beach, the equipment would then have to cross several hundred feet of shoreline to reach the work area beyond the point.

"Oh geez, not again!"

I just didn't have the heart to go through this another time: The removal of plantings, protecting the trees, restoring the beach... and probably a new driveway. There had to be another acceptable solution. If one is willing to search diligently enough (and pay the price), a solution is usually available. This turned out to be an example of... "if there's a will, there's a way," a maxim often expressed by my mother.

We even looked at coming down onto the beach over a quarter mile to the northeast and running along the shoreline. This wasn't going to fly with the soft sand, and my neighbors would have had a fit. Once in, once out, maybe, but constant trucking back and forth just was not going to go over without a major uproar–and I would have felt alike if I had been in their place.

I might as well face it. The only way to do this was to create a temporary road through the forest and swamp of 15 and 16.

To ensure the stability of the shoreline ridge was I to be forced paradoxically into changing forever the character and appearance of this wetland? Grading a road through a swamp sufficient to sustain the heavy construction equipment loads would necessitate a very substantial base. Probing the depth of the swamp with a piece of reinforcing rod, I discovered that bedrock was less than three feet beneath the surface. Any road subgrade material would punch right down to this "foundation," squeezing out all the natural soil and plant material necessary to regenerate life, as it was.

I found a route (albeit somewhat circuitous) through which a construction road could pass with the loss of only one significant tree. From the access lane it crossed the first part of the swamp, touched on a high knoll and passed across the larger part of the swamp. As the ground began to rise onto the shoreline ridge, the route followed the trench all the way to the beach. The trench would be filled, and the construction road would rest on top. I staked the 17-foot-wide perimeter of the road-to-be. Within the swamp portion of the meandering route, Greg cut down all the tall sedge grass to make it easier to visualize.

Accepting my predilection, I brainstormed a method of protecting the swamp from permanent damage and committed myself to going through with this "final" shoreline-protection work. Rick had been out to the site and was salivating to mobilize his troops and equipment. The Department of Natural Resources representative had been out, and he and the county were aware of my intentions.

The objective at completion was to have not only protected this last 285 feet of shoreline with an extension of the dimensioned-stone seawall, but also to restore the property to the way it had been originally. A contingent benefit of the construction road would be that it could give Greg access right through the middle of the property to replace trees lost during the tornado. He could plant large pines, cedars and spruce with the tree spade in areas that could otherwise never have been reached, because the swamp was in the way.

To save the swamp (or at least make it restorable) my idea was to lay down a 17-foot-wide mat of filter cloth on top of which would be dumped the rock roadbed to a level above that of the swamp. After construction was completed, the contractor would peel the rock bed off the filter cloth and haul it away. The filter cloth would be removed leaving the underlying soil, containing the root systems of the squashed vegetation, to hopefully regenerate.

Though Rick thought I had lost my marbles going to such extremes… "to save a swamp," he was more than willing to comply—with additional compensation, of course. My goal wasn't to have spent the money to buy the land, only to turn around and destroy its predominant natural feature. "Rolling up" the road as they exited the work area may seem extreme, but it was an environmentally sound answer to my desire to maintain the diversity of the natural features, which gave me the joy of being here. When it was all said and done, the extra work cost $2500.

Having placed myself in the ironic position of having to do what I was always fearful of someone else doing—destroying the character of this land, I was happy to have devised this neat solution. It was paramount to my continued enjoyment. Otherwise, I would look at what I had done with remorse forever.

Prior to installation of the temporary road, intuition warned, "have the route topographically surveyed." Fortunately, I followed through. The survey proved the old drainage trench ran uphill towards the lake, and secondly that the bottom of the trench within 20 feet of the beach was nearly identical in elevation to the surface of the access road. The caretaker and former owner of my neighbor's house complained bitterly to the town chairman and the county that I was altering the drainage patterns by filling the ditches he had dug many years ago. Ah-hah, so that's who the "trencher" was! He didn't have much credibility with the county administrators, and the town chairman did not want to get involved with the dispute.

I had explained the conditions to the county folks, illustrating with the topography that what I planned would not affect the area drainage whatsoever. Moreover, the gap in the shoreline ridge, where the trench emptied onto the beach, had been filled a year or two prior with the express consent of the DNR.

There was no convincing the former neighbor, cum caretaker, that there was nothing anyone could do to alleviate the occasional flooding in this area. During spring's heavy snowmelt and rains, the creek (which is the natural drainage course of the entire area) is of insufficient capacity to carry the load. The impounded runoff backs up all over the area. The size of the culverts, through which the creek flows under the

Drive, only makes a bad situation worse by further restricting the downstream flow. I'd make a bet that when the culverts were installed years ago (at the time the Drive was rerouted), they were sized without benefit of any engineering.

It was heart-stopping to see the transformation of this relatively pristine area into an ugly avenue of construction traffic. Load after load of limestone rock, the size of softballs, was dumped onto the filter cloth, and the trench was filled with sand and gravel out to the beach. The beach itself became the staging area for the dimensioned stone. A two-week job became a two-month job as the rock face being blasted out of the quarry supposedly was not splitting off the way it ought to. The quarry man was angry at the waste, and Rick was frustrated because they were getting so little usable stone. The delivery of stone fell way behind his crew's capability to construct the wall. So work proceeded haltingly. To this day, I'm not sure of the veracity of this ongoing excuse for the stall. It all came from Rick as I relentlessly badgered him over lack of progress.

By this time (the third time Rick's firm was doing this), I had come to know several of the workers—and they knew me and my idiosyncrasies, which they did cater to. Rick's younger brother operated the shovel, expertly lifting and nudging the large stones into place.

Winter was around the corner, and the wall had to get done before frozen ground made removal of the construction road through the swamp impossible. Greg had the final grading to do and the transplanting of thirteen large trees on higher ground between the swamp and the lake. As Thanksgiving passed, completion became problematic with the night temperatures dropping well below freezing. Fortunately, the days stayed warm and sunny—and it didn't snow.

It was with great relief that the last rock was put in place and the heavy equipment backed out the temporary road. The final grading was done to conform to the natural topography, and the trees were transplanted. Removal of the rock road was easier in theory than practice, as the shovel had to operate behind itself turning 180 degrees to load the trucks. Several laborers removed by hand rocks the shovel could not get at. With water up to their elbows, they lifted out unseen rocks they were tripping over with their rubber boots. Finally, they rolled up the sodden, tattered filter cloth. The spade came back and planted the last six trees just in from the access lane.

The finished product looked gratifyingly wonderful though there was no longer any sedge grass where the temporary road had crossed the swamp. Only time would tell if the filter cloth had done its job. The large transplanted coniferous trees put color and variety back into a forest, which had become mainly ash and birch after the tornado. The following summer only two or three shoots of sedge grass poked-up through the surface of what had become an open pond. Two wood ducks and several mallards called the pond home.

By now, several years later, much of the open pond has filled in with sedge grass, though not as thickly as it was originally. Hopefully, it just needs more time. The transplanted trees took hold and thrived as though they'd always been there. Two smaller cedars planted on the rise in the middle of the swamp became victims of a prolonged emersion in spring's flooding. Several aspens have been added, as well as a sugar maple or two on higher ground. It's quite a thrill to watch nature take over the healing process on the wounds we so carefully dressed. Ever so subtly it has woven all the threads of forest life back together to finish the job far more effectively than we ever could.

## Chapter 17
# Trees

Trees have loomed so large in this love affair. Initially I saw them as a primary attraction to this land. Since then I have become acquainted with them (each of them), smelling their heady fragrances, touching their leaves and needles, examining their buds with joy and their visible sicknesses and stress with parental concern. These giants of nature (regardless of species), with potential life spans often measured in hundreds of years, evoke within some of us an elemental feeling of being in touch with something immeasurably bigger than we are; the awesome grandeur of the natural world. In others, the trees draw forth only the desire to cut them down, whether for fun, satisfying some pioneering instinct or for commerce in terms of board-feet.

When I walk through the forest and discover piles of more than six hundred large cedar trees, I get depressed with the unnecessary loss in this area. The previous month these trees had been thriving in their environment, undisturbed for over one hundred years; a place now scarred and rutted with tracks of bulldozers and log haulers. A friend walking with me lamented, "This is a lot of death."

To put this in a practical perspective, I must remind myself that it isn't my property. On my land I have chosen other goals. I do believe that ever more people feel as I do, and eventually this ethic will predominate. Tree cutting will become as anathematized as throwing our garbage into Lake Michigan is today.

Tree planting has become a major avocation of mine. Over the course of the last several years, we have been losing the white (or paper) birch trees, ten or fifteen per year, and this last year twenty-nine. It all seemed to start with the two consecutive years of drought which may have been particularly stressful to this species. On the other hand, I have observed this happening all over northern Wisconsin, Michigan and even those areas of Ontario we recently sailed through. The affliction killing the birch trees initially becomes evident in the high branches, the foliage being sparse and the leaves never maturing in size. These trees lose their leaves far too early in the autumn, and some become barren even in late summer. No one seems to have a definitive answer as to cause. For every one I lose, I plant one. This lose one–plant one has become a way of life... but twenty-nine at one time is stretching things.

One for one is a reasonable measure, except I haven't stopped at this. My compulsion to plant coniferous trees gets the better of my senses each fall. As I wander through the woods and over the dunes, I always see another spot that would be ideal for a white pine, or whatever. So I make my list and stake the locations for Greg to transplant the largest trees they can muscle-in.

Certain birds like the dead birch for nesting. Though I would like to accommodate them, leaving the dead trees standing more than a year or two threatens neighboring trees as the progression of rot is so rapid that wind storms soon topple them. We take great care to fell the dead birch using whatever techniques necessary to avoid damaging other trees, even to the extent of cutting them down branch by branch. The dead ones are marked in late summer before the leaves fall, and the cutting takes place in early winter when the ground has frozen.

One December day I got a call from Kevin, Greg's superintendent, who with hesitant, quavering voice, informed me that a 20-inch-thick birch had gotten away from them as it fell. It was in amongst several large white pines we had transplanted. Of course, the reason for the phone call was that the falling trunk had broken off all the branches on one side of a 30-foot-high white pine.

What can you say after having emphasized and reemphasized care when cutting. The damage was done, and no amount of histrionics will put the twenty branches back on the pine tree. I did let Greg know that it doesn't do much good to plant trees, only to damage them in the operation of removing dead trees, while trying to prevent damage to healthy ones. I chided him for having Kevin call me. Chagrined, he said he had been so disgusted that he thought he'd let the culprits taste my wrath. He offered me another like-size white pine, which was fine, except it couldn't replace the damaged one on the far side of the swamp. This had been transplanted when the temporary construction road had offered an avenue for the tree spade into this otherwise totally inaccessible area of the woods. Since this incident, Greg has subcontracted the removal of the trickiest trees (those where the probability of damaging others is very high) to "specialists." The world is filled with "specialists" nowadays. Thankfully, there have been no more mishaps, at least of a magnitude that would have to be brought to my attention, before I brought it to theirs.

I do not remove birch snags (or broken-off trunks) as these are frequently used by birds. One year we had two woodpecker nests in cavities in these snags.

Woodpeckers are fun to have around and we hear their rat-a-tat-tat during every season. Often their presence is noted from a concentration of wood chips at the base of a tree. The only time their relentless hammering becomes a nuisance is when one of them has sensed a meal under the cedar shingles. It drills through the multiple layers of shingling with such precision that the integrity of the paperlike wind barrier on top of the plywood sheathing is not broken. That's remarkable control for the rapid fire of a woodpecker's beak.

Just the other day it sounded like an air hammer was puncturing the shingles high on the front wall, although the tone of the sound was heavier and louder than usual. Opening the bay window in the bedroom, I leaned out to get a better view of the wall. There on the very peak of the roof was a flicker, not the least disturbed by my loud remonstrances meant to frighten him away. The only other time I heard this quality of staccato hammering, it had been a pileated woodpecker, a very large and rather rare species of this industrious bird. It had stayed around only a day or two, which presented a dichotomy of feelings. Joy at the privilege of seeing this rare creature, yet relief that he wasn't around on a continuing basis, riddling the cedar shakes.

In the autumn, sugar maples are such brilliant complements to the somber greens of the coniferous trees. So I have substituted sugar maples for a few of the birch. Eight years ago we planted the first maple. I complained bitterly to Greg that this 12-foot tree was a sorry specimen. It had only four or five short branches. He convinced me to give it a chance. Its growth in height and fullness has been phenomenal. Each year after all the other hardwood trees have lost their leaves, this "sorry specimen," visible outside the dining room bay, is only beginning to change color. Its abundant, flaming yellow foliage is a stark contrast to an otherwise leafless forest, a demonstration of its immutable heartiness.

The year following the completion of the house, numerous

indignous cedar trees became noticeably afflicted with leaf minor, an organism that enters the leafy needles at the ends and works upward. The majority of the foliage had brown ends, and the condition of entire trees was poor. Instead of the vibrant emerald greens, the cedars were gray with little new growth to make up for the normal loss. The needles on the bottom branches of several of the spruce trees were turning rust, and the platter-sized leaves on the venerable oak tree, just off the deck, were pock-marked with dried-up circles, and were curled up like arthritic hands.

The tree doctor said it was too far along in the summer to prescribe any antidotes. Leaves were sent to the state laboratory to confirm the diagnosis. The following spring Greg began a sequenced program of spraying—initially the oak tree, prior to any leafing out, then the spruce trees for their malady, and finally four applications of insecticide phased over the summer to rid the cedars of leaf minor. The insecticide was mixed with fertilizer and sprayed over the foliage of all the cedars to simultaneously give them a systemic boost.

The stress was apparent on all the trees. They had had more than their share—the double whammy of the tornado, the resultant change in environment and then the drought. I spent those two rainless summers watering trees. I purchased 200 feet of garden hose with the intent of soaking the transplanted cedars, pines and spruce near the house. It was a difficult decision where to stop. Of all the trees needing help, I felt the transplants would be suffering the greatest stress. The other trees away from the house would, as a practical matter, have to fend for themselves, seeking whatever residual moisture they could from the sandy soil. Greg offered to come to the rescue, continuing the program on the transplants distant from the house. On a daily basis, he relocated the sprinklers and hoses (bringing over another 200-foot hose to be attached to the other hose bib on the house) three times a day, ultimately to reach the majority of the trees, regardless of their distance from the water source.

I concentrated on the trees and evergreens (junipers) around the house, setting and resetting the sprinklers to glean as much benefit as possible from each placement. This became somewhat laborious. After I dragged the hose around and set the sprinkler, I had to go all the way back to the house to turn on the spigot, only to find out the sprinkler's range wasn't reaching a tree or two it could with an adjustment of its location. To avoid repetitive treks back and forth from sprinkler to hose bib, I became very nimble at dashing to the sprinkler on the back side of the spray, moving it, and scooting out as the water came back to chase me away.

In late June and early July, this operation brought the mosquitos out of a hatch or two, so watering was made more frenetic, attempting to keep ahead of these pests. Not only did the trees begin to revive, but also the plants under them. The raspberry and thimbleberry bushes grew to become ubiquitous giants. With their large leaves, these plants dominate everywhere they grow. I had great empathy for the landscapers who were doing their part of the soaking; thrashing their way through the dense undergrowth, manipulating unwieldy hoses that have a tendency to kink. Not much fun to be searching the length of the hose in the briers looking for the damn kink.

This watering operation didn't begin until there was no detectable moisture left in the soil. Once underway it continued for many weeks, there being no respite from the sky. Whether it provided the bridge to an eventual rainstorm or not is very subjective, but we did not lose one transplanted tree. This meant a lot to me, as dead transplanted trees could not be replaced in size. The spade had planted them in sequence from farthest in to last out of the planting lanes. This is to say nothing of what could have been the lost investment and the heartbreak of failure, after having put so much effort into recreating the tornado-decimated woods.

\* \* \* \*

Lake Michigan's high-water years had caused a commensurate increase in the water level of the creek, drowning all that was growing in its floodplain from the beach inland to the Drive. With the drought came the reappearance of the floodplain and regermination of grasses and wildflowers along the creek, but not the trees. Looking downstream from the Drive, the flanking high ground had lost most of its conifer green in the tornado. Yet, it remained a scenic view, with the horizon on Lake Michigan the top side of a shallow triangle of blue, seen through the gap. Somewhat against our better judgement, we decided to chance the reoccurrence of the record-breaking high water and replant the floodplain with several cedars. These trees seem to do well in somewhat wet soil and lots of sunshine, the precise environment found along the creek. Should ever the lake level rise to those records again, I guess these thriving, verdant cedars will have to be moved.

That autumn someone mutilated several of the trees and bushes along the top of the floodplain embankment. The vandal had cut off, at chest height, several small cedars and a blue spruce we had planted, along with all the dogwood and alders that obstructed the view of the creek and its floodplain from the high ground. A beautiful spruce tree clinging to the side of the bank had survived all of nature's onslaughts, only to have its top ten feet sliced off by this jerk. The why of this purposeful destruction made no sense. It took a lot of work. It wasn't done with a chain saw, but sawn by hand and the bushes "trimmed" with clippers. We found a large stump in the woods that had been made into a seat from which it was possible to view the creek. Could it have been a bow hunter trying to create a broad field of fire into this area frequented by deer? Even this didn't add up because of the extent of damage.

As I stood there discovering the damage tree by tree, my feelings were akin to those experienced after the tornado, with the added emotion of anger directed at this thoughtless destruction of another's property. I reported the vandalism to the sheriff, and an investigator came out. He made out a report, but shed no more light on probable cause then could we. To make things even more bizarre, the vandal (for lack of a better term) not only trimmed off the trees and bushes, but cut away the sod and earth on top of the culvert through which the creek ran under the Drive.

Greg turned up a branch on each of the spruce trees, lashing them to the trunks as leaders to eventually become new trunks. He also gave me a price to replace the damaged trees, which I used for an insurance claim and subsequent replacement. The reasoning behind the dirty deed remains a mystery.

Tamarack trees were once common in Door County, but were logged off for use as pilings. These conifers, whose foliage turns yellow in the fall before dropping their needles, have always been a favorite of mine. I've planted eleven. After struggling mightily for two years to reach above the competing tall grasses, they finally took off in a spurt of sun-drenched growth, sprouting annually in excess of two feet. Now at six-to-eight-feet-high they have become posts on which the deer rub their antlers in the fall. We've lost two, which were girdled in this manner. Inexplicably, the deer never seem to use the birch for rubbing posts, favoring only the poplars and tamaracks.

To facilitate building the house, Don had cut down a large clump of scrubby trees, which had black, cigar-like growths all over their branches and twigs. It never occurred to me to inquire about these ugly growths, yet a year or so later, these ulcerations were appearing on trees elsewhere. I asked Greg what this was and learned it was a disease endemic to cherry trees. It spreads easily from tree to tree. He called it black knot and stated that there was no cure.

My attention having been drawn to the wild cherry trees, I was astounded at the abundance of this species on the property, from seedlings to several beautiful specimens over 40 feet tall. For three or four days in the spring, the crowns of the larger cherry trees are suffused with blossoms. Great puffs of white, high in the forest canopy, softly complement the permanent background greens of the cedar and pine. The cherry blossoms don't last long, but while present are not only beautiful to look at, but sweeten the springtime redolence of the forest with a delicate bouquet.

The larger cherry trees were free of black knot and I was determined that they remain so. It was simply a matter of cutting away infected branches and stems on the smaller trees and bagging them to be hauled away and burned offsite. So with my clippers and bushel basket I began my campaign to eradicate black knot. It has always been a late fall or early winter activity. The leaves need to have dropped so the growths can be seen. The temperature has to be cold enough to make it comfortable to work, yet the forest floor has to be free of snow, as many of the black knots are within a few inches of the ground. Seeing an infestation I clip and saw away, filling bushel basket after bushel basket with the infected wood. When gathered in heavy concentrations in the basket, the black knot gives off a particularly pungent and foul odor. Over the years I have succeeded in eliminating most of the visible evidence of this disease. But each fall there's always more.

Poplars spurted in growth, subsequent to the tornado, reaching phenomenal heights in proportion to their skinny girths. Each summer I note that several poplars seem to suffer destruction or mutilation of their topmost branches. Either the small branches have been snapped off close to the trunk, or are hanging lifeless, broken and stripped of leaves and twigs. The tree doctor and I remained baffled as to cause until one summer evening, when sitting in the window seat of the big bay, I spied a porcupine at the very top of a poplar's slender trunk. He reached out and bent back branch after branch till they cracked. Then he munched away on the tender twigs and leaves, diligently stripping the tree bare of its uppermost foliage.

The trees subjected to this abuse become virtual wrecks, yet there's no solution without eliminating the porcupines, a Hitlerian answer, totally foreign in concept to my makeup. It is the amalgamation of all the life within the forest that gives it its persona. In the broader perspective, the absence of this animal in the network of forest life would in someway impoverish the whole. The porcupine is more of a concern to me from the standpoint of a chance encounter with Muggs. One night in bed, Muggs was worrying a paw to the exclusion of sleep, which is rare indeed. I discovered a quill imbedded in a paw pad. I plucked it out so fast that Muggs didn't have time to think about it. The quill was much sharper than a needle. I had visions of dozens of these vicious weapons stuck in her nose.

She had escaped a close encounter with a porcupine who ambled (their top speed is an amble) along the perimeter of the deck. I was reading, and Muggs was lying on her side on the deck, fast asleep after a long walk on the beach. A porcupine appeared from under the junipers to the left of the deck and in a typically deliberate manner, slowly strolled across the stone walk and then along the face of the raised deck, not two feet in front of Muggs' nose. It was as oblivious to the presence of either of us, as Muggs was to this creature's invasion of her territory. I felt like some giant hand was squeezing my insides. I couldn't walk over to Muggs to hold on to her collar for fear I'd awaken her in my attempt to protect her. Yet not doing so was as perilous. I was almost afraid to breathe. Time seemed to stand still until the ponderous creature perambulated beyond the deck, disappearing into the large thicket of yews north of the oak tree. It was only then that I recovered sufficiently to go over and snap the leash on sleeping beauty, who instantly was wide awake and back on watch. Porcupines and skunks are cause for special concern.

It was incredible that Muggs didn't catch the scent. Often when I am sitting on the deck with her, she will suddenly become alert to something unseen approaching from behind the point, well over a hundred yards away. She sticks her nose toward the direction from which the breeze is coming, wrinkles it up and weaves back and forth with low growls emanating from the depths of her throat. Inevitably, several minutes later someone will come into sight along the beach. She is a fail-safe early warning system, detecting the downwind approach of visitors, human or animal.

\* \* \* \*

The Drive is one of Wisconsin's designated "rustic roads," an encomium bestowed upon a few unspoiled scenic byways across the state by a department of the state government that sees to these sort of things. It must be a lonely life for the rustic road designators, with so many of their bureaucratic brethren diligently working on policies and directives with effects antipodal to their concerns. No sooner than an element of our natural world receives a measure of protection, it seems someone is contemplating measures which would strip the resource of its preservation value.

Wildflowers grow in great abundance and diversity along the narrow shoulders of the Drive, a colorful seasonal feature of the rustic road (and many other country roads). Typically they become victims of the highway department mowers, operated by the same guys who plow the snow—ever wider off the pavement. Ergo, gone are the colorful verges with their treasure of wildflowers which are so much enjoyed. In fact, the shoulder mowing became a hot controversy and, figuratively, took an act of Congress to get the mowing stopped or just reduced in frequency. Its resolution required hearings and expert testimony.

If the county's policies and attitudes can be labeled as Neanderthal, the Wisconsin Public Service Company is positively paleolithic in environmental consciousness. What makes this organization's policies even scarier than government's is that recourse or enlightenment is even more frustrating than with government. At least government is answerable to the people. This organization's insensitive tree trimming policies could be termed the measure of mutilation and destruction of trees.

They inform the property owners along the right-of-way that tree-trimming is about to take place and describe their standards of trimming (which are extreme). If it's on private property you have a chance to stop it, but not along the county right-of-way.

I read with dismay their letter of intent. When next we returned (only a few days later) the deed had already been done. They had fulfilled their commitment of notification and proceeded posthaste. The results were grotesque. Where there had been a balanced canopy over the Drive, the cutting had been scythe-like in aesthetic butchery along the wireway. One-sided trees were created and other large cedar tree trunks lopped off, leaving only a branch or two remaining. Trees were cut down that would have taken a generation of growth before they would have interfered with the power lines, a crude cutback that left poles and wires the predominant visual feature along this rustic road. Piles of cutup trees, which they never bothered to return for, remain a mute testament to this senseless overkill dictated by a single-minded public utility.

A quarter mile down the Drive in the direction of town, the ownership of the transmission lines changes from Wisconsin Public Service Company to the Sturgeon Bay Combined Utilities. This company undertook a coincident trimming program. At the point along the

Drive where the two meet, the contrast in trimming was as stark as night and day. Sturgeon Bay Utilities cutback was so artfully accomplished as to be almost unnoticeable along their entire stretch of roadway. Their trimming may have to be repeated at a greater frequency, but on the other hand, it probably took half the time.

The harsh contrast between the trimming these two utilities did remains depressingly visible along the roadway. The author is not alone disputing utility companys' tree-trimming policies. More than one citizen organization has as a goal stopping this needless environmental destruction.

\* \* \* \*

The triangle of land formed by the junction of the Drive and the access lane is not visible from the north on the road, but is conversely front and center for several hundred feet when approaching from the south. The view of this was not particularly pleasant. The larger trees in this area had fallen, and there was an old, rotten signpost frame standing in front of a scrubby thicket of alders and poplars. The grade dropped quickly off the shoulders into a low area, which was an intermittent watercourse. In times of heavy rain and flooding, it provided an outlet to the backed up water through a small culvert, under the gravel access road, from which the watercourse led eventually back to the creek.

This thicket had always bothered me and, in fact, had somewhat dissuaded me when we were looking for land and considering this as THE place. I'm sure it had also been a turnoff to other folks snooping around for property. This triangle is located at one end of a huge (twelve-acre) football-shaped parcel of land, flanked on one side by the Drive and on the other by the gravel lane, which emanates back onto the Drive a quarter-mile northeast.

Though the twelve acres had been quite badly devastated by the tornado, its least-affected portion was adjacent to my property.

Having now purchased twice from George, we had gotten to know each other better, and he knew that I knew he wanted out. It only took a phone call to make him aware that I wanted to buy another of his "white elephants." This one wasn't really a swamp, but was subject to seasonal flooding from the meandering creek that flowed through it. I purchased that part of the "football" contiguous with my land on the other side of the access lane.

Greg and I got together to discuss my idea of constructing a dry-laid weatheredge limestone retaining wall in the deepest part of the drainage swale. The scrubby vegetation would be removed in the area between the swale and the junction of the pavements. This entire area would be filled and large coniferous trees planted throughout the triangle, both above and beneath the wall. The trees would hopefully create a more scenic view for the passerby going up the Drive and would help frame the entry to the access lane, a counterpoint to the spruce trees on the other side.

After the stone wall was in place I selected thirteen cedars and spruce on property Greg owned and four pines on Dennis' property, all of which could be reached by the large spade (in dry weather). We staked the location of each tree in the triangle to create a somewhat pyramidal combination of trees to view when approaching from the southwest. I could hardly wait to get back to see the results. When I made the last bend in the road, the scene out the windshield revealed a spectacular difference. The nondescript, scrubby thicket at the junction had been transformed into a beautiful copse of evergreen trees of staggered height and varied shades of green. A tall (35-foot) white pine towered over all from behind the cedars. At the apex, the verdure was protected from snowplows by several large wood posts.

Now I had the ability to properly place the Schloss' fire number plaque. I had had two smaller plaques made with white arrows, one straight and one hooked, on matching green backgrounds. I placed both my neighbor's fire number and mine behind adjacent wood posts and underneath the numbers bolted on the directional arrow plaques, his hooked and mine pointing to the right. It was the last time people would come to the wrong house, which was a blessing because the neighbor's house was for sale at the time.

A year later someone took a fancy to the hooked arrow and unbolted it from the iron post, leaving me the nuts and bolts. This neat little arrow probably ended up in one of those bedroom museums, as did the red fire number. I had another made, and when I bolted it on I distended the bolt ends so this time they'll have to cut it off with a hacksaw.

## Chapter 18

# Summer

As with many things, even the most inviting or delectable have a tendency to lose their luster if they become too much of a good thing. Spring is no exception when it lingers too long. A bright sunny day of 50 degrees in early April transmits a totally different image to our psyches than does a day of the same characteristic in early June. Spring has a difficult time giving up to summer along the shoreline. Inland a half-mile the trees have developed their full canopies of leaves, and temperatures warrant short sleeves. Yet the littoral's natural world stays seasonally retarded by refrigerated onshore breezes and clammy fog. Sunlight still streams, almost full, through the lacy canopy of delicate green overhead. Trillium remains in protracted bloom. The leaves of the venerable oak and sugar maples stay in suspended immaturity awaiting, week after week, the promise of summer warmth and sunshine, unfiltered by the gossamer-like curtain of moisture pervasive in the atmosphere over the lake. The only truly clear days seem to be those accompanying high pressure, knifing the last blasts of cold air down from the Arctic. Though inland folks scurry about protecting their emerging gardens from unseasonable frost, we along the shore stay warmer with the lake's stable temperature moderating the early morning's freeze. These are nice times to walk along the beach buffered by the dunes and the forest from the bite of the northwest wind. It is at this time of year I have most frequently heard the call of loons. Valuing their privacy they never come close to the beach, so binoculars are essential to get a good look at these emblematic creatures of the wilderness. I don't know of anyone whose primal senses are not deeply touched when hearing the weird and tremulous call of a loon.

Much closer to view, while recumbent on the window seat sipping an iced tea, I saw a flash of orange cross the meadow and light upon a branch of a birch; and then another orange bird darted by the windows. This was too much to miss, even if I had to arouse myself from the lassitude induced by a day working on *Aurora*. The binoculars were focused on a pair of orioles and, unbelievably, I caught sight of a scarlet tanager in the oak tree. Never before, nor since, have I seen these birds, though admittedly I haven't looked very hard.

Summertime pleasures in the out-of-doors are of a more intimate, closer-at-hand sort than those of the other seasons. It's hard to equate intimacy (which connotates warmth) with winter–period! Nor with autumn's brash, brassy flamboyancy. Springtime, though nature's new year, is yet a season of extended horizons. Still the see-through forest, it only tantalizes us with a stroke of delicate color here and there, a fleeting caress of warmth and hints of the myriad life-forms soon to be enjoyed right before our noses during summer.

Except for an infrequent season of little rain, summer's mosquitos preclude my walks in the forest. It may not bother some people (I often wonder how surveyors keep from going mad), but the clouds of these pests take the fun out of the forest for me, at least during the month of June. There's no lack of things I like to do around the Schloss if there is a sufficient breeze to keep the mosquitos off the beachfront.

Some folks play golf, others tennis. When I'm not sailing, I like to tend my landscaping, weeding being a major endeavor. The junipers around the house and on the meadow have grown like wildfire since their planting eight years ago. They now form large carpeted areas, which require ongoing attention to keep them free of competing grasses growing up through the evergreens. These stems I pull out, roots and all. Weeding the junipers for the first few years was quite time-consuming, but eventually I got ahead of the game and now it's only an hour or two every couple of weeks. Once while weeding the junipers in front of the house, I noticed a fox pop out of the forest and saunter directly toward me across the meadow. He must have been singing to himself "what a wonderful day," because he was asleep at the switch, not looking where he was going. We all have our moments. On my hands and knees we were just about at the same level, and he got to within six feet before he woke up. Eyeball to eyeball we stared each other down. Neither moved a muscle until I made a sound to break the spell, and the fox leaped for the forest, tail streaming behind. It must have been the same fox whose rounds for a long while took him along the walk behind the house, daily at about 5:30 p.m. Jan was usually fixing dinner at this time, so noticed him when he passed the kitchen window and missed him when he didn't show up anymore.

Maybe he was one of the previous year's kits, raised in a den under the stump of an uprooted pine alongside the path out to the point. It had been fun watching the fox family grow up. At first, the only evidence of habitation was leftover hair and bone outside one of the entries. Then several pairs of eyes peered out from the dark recesses, sometimes one pair on top of another; silently watching their world-to-be, timorously awaiting the return of their mother. Soon noses tentatively poked out into the sunlight, and the boldest lay quietly on the warm earth just outside the hole. Eventually curiosity overcame timidity, bringing all four out into their yard to play and tussle, until alarmed by something or other, they'd dash for the safety of the den. Slowly the eyes and noses would reappear as they regained their courage. And then one day we looked and the den was empty, with only the meal scraps strewn about to remind us of the wonderful creatures whose home had been the labyrinth under the stump.

Muggs is ever curious about holes in the ground and caverns under stumps, but her inquisitiveness has been tempered with caution. We liked walking along a deer trail into the forest a short distance from the Schloss. Its attraction for me was the scenery and for her, all the scents of other creatures. Irresistibly, she had her head and shoulders down into a large opening, under the root wheel of an upended tree, when she was greeted with a growl of sorts from within the hollow. Her broad rear end and upturned tail retreated out as fast as her short legs would operate in reverse. She is still curious about holes, but only from the outside, and since her close encounter always balks at walking along that particular path.

Diligent clearing of the broadleaf weeds from the meadow has promoted the return of the wide variety of tall grass species, which were waving in the breeze when I first came upon the scene. Never changing has been the profusion of delicate pink wild roses in bloom where the meadow meets the forest. Behind the house I've let the grasses and weeds do their thing once clear of the junipers on top of the stone wall. This area nurtures a wonderful variety of wildflowers. Columbine, orange hawkweed, harebells and phlox provide a sprinkling of color throughout the summer season in sunny patches in the woods. Not knowing much about wildflowers, we bought a field guide to help identify these delicate plants, which add so much to enjoying summer in the woods and along the shore. It was quite thrilling to discover jack-in-the-pulpits and nodding trillium growing secretly in shady areas, their blooms hidden by their leaves, and the flowers themselves all but concealed by other leafy plants on the forest floor. The buttercups and marsh marigolds of the spring swamp have been replaced by the dramatic blue flag iris. We've identified and catalogued forty different species of wildflowers on the seven acres.

The thickets of thimbleberries and raspberries, so prolific and giant-like on the forest floor, and the low leaf canopy above, conceal all animal life from view. So much is going on unseen so close. Only occasionally do we get a glimpse of life within the forest. As with the fox, my quiet avocation of tending to the landscaping has blessed me with other close encounters with the creatures of the forest. One afternoon a small fawn popped out of the edge of the forest and headed toward me, seemingly without fear. When it got to within a few feet, I moved, and it stopped, wondering I suppose, what do I do now? After several minutes of studying each other, it turned and walked with delicate step back through the curtain of foliage with a parting glance over its shoulder; truly an innocent abroad.

Seeing deer has of recent years become commonplace even in the backyards of suburbia. These beautiful animals in their quest for nourishment in shrinking habitats have become so prevalent as to be almost ubiquitous in areas where they were unheard of twenty years ago, an at-home vignette of what is happening to wildlife throughout the world. An inexorable competition between deer and development, elephants and cattle farms, tigers, bears and magic potions and owls and salmon versus logging. The list could continue for pages in staggering dimension—an apocalyptic cataloging of a natural world being shorn of its diversity; a vision of environmental collapse.

Deer are an anomaly in the grim census of species population losses. They've adapted and, without natural predation, have grown to become a controversial nuisance in places, despite the legions of hunters thrilled by the blood sport of "getting their buck" each November. Somehow baiting the animals with corn or apples for weeks ahead of the season and then sitting hidden in a tree stand over the bait to blast the animal to oblivion as it comes to breakfast doesn't seem very sporting to me.

Despite their becoming commonplace in our lives, the deer remain mostly out of sight, but when nearby and in view, they help keep us in touch with the natural world. Arriving on a summer evening, we've surprised them lying in the grassy meadow in front of the house, apparently less annoyed by the mosquitos there than in the forest. On an early summer morning, before humankind is aware that a new day has dawned, I've awakened to a splashing sound on the beach. Getting up for other purposes I chanced a look at the shore and saw six deer frolicking in the shallows, rearing on their hind legs with the water cascading down their backs, lunging out, up on the sand and playfully prancing about. This lasted for fifteen or twenty minutes until, at some signal from one not engaged in the games, they abandoned the play and quickly trotted into the protective forest, followed by the lookout. At the same time of day the bathroom window has framed deer just outside. One year they became fond of the hosta lilies, but the growth of the hostas was greater than the appetites of the deer.

One day while having lunch at the table in the small bay facing the oak tree, a very poignant drama unfolded before my eyes. Beyond the oak a fawn emerged from the trees and underbrush beside the path leading to the beach. It was typically tentative, yet its demeanor was unusual. Purposeful and searching, it trotted on the path to the top of the dune to look down the driveway. Then it hurriedly returned and stood glancing about like a sentry on duty on the edge of no-man's-land. Just as it disappeared back into the forest, out from behind the curtain 70 or 80 feet down the beach popped another fawn, repeating the same drill, but in a different direction, before returning to the cover of the forest. Though gone from sight, they still had my quizzical attention, sensing that something out of the ordinary was taking place. And it was. Within a short time, led by one fawn and followed by the other protectively close behind, a doe hobbled slowly and directly across the beachfront meadow. Each step of the mother an agony, using the stub end of half a right front leg to balance herself for the next step onward. Up and down her awkward pace was repeated until the family disappeared one by one in line of march into the flanking forest. Although their range was extremely constricted by the doe's crippled condition, I never saw them again. I hope they somehow understood that this forest offered them a sanctuary.

Another strange and inexplicable incident with deer occurred on a beautiful evening in June when a guest, looking out the bay window, said humorously, but with some question, "There's a shark."

"Sure, Jeff."

"No, I mean it. Look for yourself. It looks like a dorsal fin."

There was something moving along the water's surface. The lake was glassy calm and the horizon undiscernible—the water perfectly mirroring the sky. I recognized a deer swimming, its head the only thing visible. It wasn't far offshore, but it slowly became apparent that it was swimming directly away from us toward the unbroken horizon. The binoculars confirmed this. It seemed only unusual at first, but odd enough to keep our interest until it became horrifyingly certain that this animal was swimming with singular strength and purpose directly out to sea—almost certainly to its destiny. We watched transfixed in our helplessness as the head became a dot, and then the dot disappeared, swallowed by mere distance. Each of us who had observed this (and there were five of us) were visibly affected. At dinner immediately following, the conversation, normally boisterous, was about as quiet as the silence over the surface of the evening's calm sea. It disturbed my sleep that night and haunted me for days.

When learning of the house ("your summer cottage") on the lake, people from other parts of the country have exclaimed, "Oh, how wonderful, your own beach to relax on and swim off."

"Well, yes and no," I reply. Seeing their perplexed looks I let them know it is most delightful to drag our beach chairs down to the shore and bask in the sun's warmth, but swimming is a different thing altogether. It's not that I don't ever do it, but it is an infrequent pleasure. The water is cold, achingly so. Though the surface temperature can vary widely with differing wind directions, even on the warmest-water days of a normal summer, it takes some fortitude to make the plunge. Some do just that, "Oh hell, let's get the shock over with." Others do it inch by torturous inch, wiping water up their arms and chests to aid acclimatizing. It doesn't seem to get any easier as the water gets deeper, from calves to thighs to crotch and upwards, it's just as painful. By the time it reaches my swimsuit I've lost all my patience and throwing caution to the wind, take the plunge. But this only occurs two or three times a year and during a cool summer, maybe not at all.

My usual motivation to undergo this water torture stems from working up a sweat on the landscaping with the attendant discomfort of the biting insects. I walk out to the beach and into the lake to cool off. At these times the chill is delicious. One day I was clearing out branches and logs from the creek, an arduous task in the best of circumstances. With sweat streaming into my eyes and bugs into all the openings, I persisted until finished, having already made up my mind that my reward was a swim. I didn't even bother to remove my clothes.

Onshore breezes push the warmer surface water inshore. If the absence of wind overnight leaves it there, in the morning the lake along the shore is crystal clear, all particulate matter having precipitated out. The sand wavelets and stones on the bottom are distinct as are one's toes several feet under the surface. At these times, on a sunny day, a swim is really a welcome interlude. The sun dries me off rapidly, as I drip through the canvas chair bottom.

On breezy days with an onshore wind, the seagulls spend their time in aerobatic demonstrations above the shoreline. Catching updrafts for lift, they effortlessly wheel and dive for hours on end. Their enjoyment is raucously vocal with their screechy, oww, owws; sounding like a child whose arm is being twisted by a tormenting big brother.

It's very hard for me to read on the beach. So much is going on in the natural world around me. And if nothing is happening at a given moment, the view down the coastline and across the blue water is mesmerizing. Along with the vista, the sound and rhythm of the gentle surge of water washing up the sand is soporific.

One year two Canada geese made their home along the beach. Six goslings were herded to-and-fro all summer long, growing in size until they were hard to distinguish from the parents. They became so accustomed to our presence that their detour around us ceased to be into the water, but behind us on the beach. Unless, of course, Muggs was there watching all the goings-on from within the beach umbrella's shady ellipse, then they gave us a wide berth in the water.

Muggs' patience with the biting flies on the beach lasts only so long. When she starts trying to bite them back in flight, it's time to go inside. There's a certain type of fly (not the ordinary kind) which resembles a miniature stealth bomber. Like the airplane, you don't notice them until they are doing their damage, attacking with a particularly painful bite.

When one insidiously zeros in on an inaccessible part of my back, I react instinctively, usually managing only to give my hand a crack against the aluminum frame of the beach chair. I can sympathize with Muggs whose back and rump are unprotected targets of opportunity. Contrarily, they are easier to swat than the ordinary fly. It seems they don't have quite the same instinctive early warning system as their cousins. Or, maybe it just takes them an instant longer to release their fangs from their tenacious grip. So there is some satisfaction realized in just retribution. Why is it that flies are so attracted to beaches, wherever?

Some years for weeks at a time the dead alewives wash ashore by the thousands, stranding at the high-water mark. They are pushed higher up the beach in successive rows by higher wash on windier days. These small, oily fish are an inedible species, only good for grinding up into fertilizer. Inexplicably, they die off by the millions each summer to decay on the beaches. With the exception of their eyes, I don't even think the gulls find them very appetizing. In bad years I've tried raking them into pits I've dug in the sand, only to be frustrated a day later when, after a strong onshore wind, there are thousands of replacements. Call Greg!

His guys have dug more and bigger pits with a Bobcat, only to have a high sea wash out the pits. They've shoveled them into wheelbarrows and trucked them away, but no one wants to be downwind. The best solution seems to be raking them into piles and burning them. Anyway you do it, it is labor-intensive, expensive and undertaken only when they are of a number that make it impossible to open the beachside windows or to enjoy being in front of the house. Forget being on the beach, the flies are far more numerous than the alewives. When an onshore wind pushes the flies inland, the front of the house looks like pepper sprinkled on cream cheese.

On these days as far as the eye can see down the strand of beach, it is devoid of any human activity. How deceptive is the appearance of paradise.

As paradise is fickle, so are the breezes variable on this inland sea. Fortunately, tomorrow's wind probably will be offshore, cleansing the atmosphere of humidity, flies and fish factory smells. On the days of malodorous breezes, it's best to plan on an alternative activity.

"Tennis anyone?"

Nope, but it is fun to nose through a few of the many antique shops and barns in the county. Once in awhile I've come upon a real gem like an Indian birchbark basket, a copper kettle and a single sheave block from some ship of old, pointed out by Jan. This I restored to its original hardwood lustre and painted the iron hook black. It now hangs from the ridge beam projecting from under the front gable of the garage.

Jan has an eye for real finds and the patience to search behind what's out in front. Her collection is displayed on top of the cabinetry: copper and brass ware, candle molds, pepper grinders, salt cellars, delft vases and pitchers. An old wooden rake and a long gourd ladle live under the companionway steps to the loft. Crocks of varying sizes and shapes congregate under the eaves in the loft, and an Italian wine bottle, half as big as I am, reposes in an iron basket on the loft deck. Its imperfect green glass picks up the most unusual hues reflected from the lake on bright sunny days.

My mother's older brother was a fine artist of some repute, specializing in marines and sailing vessels. His most striking works were night scenes at sea or moonlit shorelines. Two of these grace the walls of the Schloss. Canvases, unframed, which complement the simplicity of the interior and speak graphically and eloquently of the drama unfolding before our eyes on a daily basis.

\* \* \* \*

A west wind brings uncustomary heat, tempered not in the least by the natural refrigerator in front of the Schloss. This overland breeze can rapidly push up the thermometer to 85 degrees. When the breeze begins to blow hot air through the open windows, it's time to close up the house and drop the awnings on the south side to retain within the early morning coolness. With the excellent insulation, the interior of the house in this mode stays more than ten degrees cooler than the outside temperature. Anyone walking along the beach would think the occupants were recluses. With the land heating up by early afternoon, the wind direction over the lake often backs suddenly into the south, dropping the temperature along the shore dramatically. The refraction (my term—not a meteorologist's) of the wind direction, even into the southwest, will bring the relief of cooler air along the beach. The temperature can spike up and down several times as the two breezes vie for dominance. On these days, opening and closing up becomes a game of musical windows until I lose my patience and leave the windows shut until the sun drops behind the trees in the west.

I have often left the Schloss with a jacket on to drive into Sturgeon Bay and watched the outdoor temperature register on the digital dash of my space-age car rise from the sixties to the upper eighties by the time I've reached town. If the car didn't have the thermometer the change would never occur to me, because my space-age air conditioner modulates the auto's interior temperature to a jacket comfort. It's only when reaching town and seeing people in shorts that it occurs to me that I've returned to real summertime. Getting out of the car with my jacket on, I've wondered if anyone seeing me is thinking that I'm a premature Geritol case. Since I'm not a fan of hot weather, as I approach Lake Michigan on my way back to the Schloss, it's always a delight to feel the cool air wash over my arm, elbowed out the window.

The changes in weather often come not subtly, but suddenly with dramatic and destructive ferocity. Though sunny and clear, the visibility is poor and the atmosphere oppressive with sticky humidity. The flies bite with an unusual aggressiveness. Small birds dart about frenetically. Thunder is faintly audible. Something's on the way. The barometer has been heralding the change for quite a while. From the beach I can see the tops of the towering vanguard of cumulus clouds billowing into the stratosphere in the northwest. Soon the main body appears—a definitive line of black rising low over the trees, pushing the blue sky relentlessly before it. As it nears, it casts a sickly yellow hue on daytime. The inky bottom, with a hint of green, is torn apart with tendrils of cloud mixing about under the ceiling, evidence of lots of wind aloft. As the squall line approaches there is a palpable increase in tension. Conversely there is a feeling of comfort and security in being ashore. As the line passes overhead, the wind strikes with vicious determination to rip apart anything in its path. The tall pines bend and sway to the pressure. The branches of the oak flail about wildly, leaves and twigs tearing off.

It's time to go inside and shut the windows. From the safety of the window seat it's fun to watch this high drama. The squall moves rapidly out to seaward, the wind line starkly visible, marching out across the open water. Its leading edge as well-defined on the water's surface as it was in the sky. The rain follows momentarily, obscuring everything. Driven horizontally by the wind it pelts against the house, adding another dimension to the screeching wind and thunderclaps overhead. Water pours off the valleys in torrents, and the rain barrel fills to overflowing in seconds. For minutes at a time, visibility is reduced to a few yards with only momentary glimpses of the treetops thrashing wildly about. The thunder is now to seaward, and the tempo of the storm moderates. The deluge becomes just rain and soon ends. The wind has abated, and the only noise is the water dripping off the foliage overhead. Strangely, this soft sound catches hold of our senses more than does the thunder, audible in the distance. The atmosphere is damp, but not close; the air cleansed and sweetly fragrant with the essence of cedar and spruce.

Over the water, the back side of the storm has broken up into individual clouds from which rain is streaking downward in diagonal curtains. Soon the sun has found the holes, and its rays begin to compete in the theater of rain-washed sky. When the sun touches our earth, it is also reaching the benign back side of the receding squall line. The tops of the towering clouds are bathed in incandescent light in bold sculptural contrast to the flat, shadowed grays underneath, now squeezing over the horizon. Occasionally the sky over the water hosts a rainbow. An encore to a performance of grand scale; one which never loses its capacity to awe those who have experienced it.

\* \* \* \*

Along the Lake Michigan littoral, summer hardly seems to have begun when, shockingly, we are looking down the steep back side of the season. So subtle as to be almost beyond perception, we become aware that the greens have lost their richness. Goldenrod and asters have appeared where daisies and chicory had a brief time before dominated, and blessedly, the mosquitos have disappeared.

I find more ambition to get into larger landscaping maintenance tasks that have been bugging me for a month or two. The annual flush of growth on the deciduous trees shades out some of the cedars. The rampant growth of grasses and thimbleberry bushes chokes off all growth in the lower portion of the cedars and spruce. With my polesaw I can reach in and cut out limbs and weedy trunks of mountain maple and dogwood. I leave the tougher stuff for Greg.

The poison ivy has been left alone to flourish in places along the forest verge. The first year or two the ivy was so widespread that Greg had to come in to spray with a tank on his back. The repeated spraying of the first years minimized any resurgent growth, but it still requires searching for these innocent-looking plants which can cause so much discomfort. On one of our visits to the house during construction, John looked clown-like with his arms and face plastered with a white paste. But this was a sad jester with a case of poison ivy as virulent as I've ever seen. He said he got it next to the foundation. When looking for it I discovered large patches all around the construction area. That's when I had Greg begin the spray campaign—couldn't have the carpenters missing work because they were nursing poison ivy. Now I just walk around with an aerosol spray can.

Its very directional spray reaches several feet in, under and through other desirable plants which you don't want to touch with the poison ivy killer. A white soapy coating is left on what it soaks, so you know what you've sprayed. My aim is unerring. It's hard to believe that in all these years of walking through areas where this extremely toxic plant is found, I've apparently never touched one—either that or I am immune. On the other hand, I am very careful where I step. Control of poison ivy is the sort of thing that is never 100 percent. Just when you haven't seen any in an area for a year or two there's one or two of the droopy three-leaved plants thriving in amongst other harmless vegetation.

The toughest undesirable stuff to deal with are the thimbleberries. Its growth is rampant, and it takes over wherever it is to a height of around four feet. Nothing else does very well under the canopy of platter-size leaves. This includes the majority of the foliage of smaller cedars and spruce struggling for survival. Pulling up thimbleberry stalks is not rewarding, unlike raspberry bushes which are fun to pull up. In the sandy soil the raspberries most often come up along with their long trailing root, whereas the root system of the thimbleberry is as tenacious as a bulldog. More often than not you pull on one until you think you're going to get a rupture, and then the root breaks, with the puller losing his balance and composure. "Spray 'em, Greg." It's the only way. Spraying the leaves somehow kills the whole plant, roots and all.

The tall grasses also choke out the evergreen foliage, turning it all brown. These thick intrusive grass stems I pull out around the cedars near the house, but leave the problem areas in the buggy forest for the guy with the power clipper.

When I get tired of being busy on a late-summer day, I sit contentedly on the deck, relaxed without swatting any bugs. The geraniums in the flower boxes are as big as football mums. Hummingbirds hover from bloom to bloom extracting the nectar with their needle-like beaks. Butterflies float leisurely about, also feasting upon the abundant flowered ambrosia dotting the meadow. The monarchs seem to especially savor the milkweed, which must be to them like wine is to Frenchmen. Each to his own. My weakness is ice-cream sundaes. The blooms of the fleshy sedum along the side of the deck have finally turned from green to pink to so remain until frost changes the pink to a coppery brown. Several years ago, the sedum was cut and dried for an arrangement in a large shallow copper vessel. We must have done something right, because these dried flowers have adorned the gateleg table in the small bay ever since.

Looking out the dining area bay three cedar trees are a deep emerald green background to a large and prolific bed of lilies. These vivid yellow and orange flowers, on long willowy stems swinging in the breeze, remind me of "the wave" at a college football game.

The clusters of seeds on the tall grasses have turned to color the meadow with a soft wash of mauve. Even the air has somehow changed. I sense a drier and sharper fragrance close to the pines and spruce; the carpet of pine needles on the forest floor is especially pungent on sunny days.

Starlit nights of late summer can be an extraordinary treat. To sit on the sand or walk along the shore without the annoyance of bugs, still enjoying only the coolness that comes with a summer night—not bracing or chill, but simply a soft contrast to the daytime heat. Without a moon, it is so dark that walking has the concern of blindness, yet the indigo sky is filled with a zillion stars. Up and down the coastline no ambient glow from sodium vapor looms interrupt the infinity of night over the lake. Only the moon can do that with its lambent glow playing lightly over cat's paws of breeze shimmering on an otherwise calm sea. Between us and the moon occasional clouds seem to be suspended in this lofty scene. Our surroundings darken noticeably, if only momentarily, as one drifts languidly across the moon. By contrast, the sky behind has a diffused glow and the backlit cloud its silver lining.

Though drowsy, it's hard to break away from the reverie of these nights on the beach, rare as they are. Why go in, we're not cold, and we don't have to get up in the morning. Responsibility and commitment are of another world. This is a time for lovers. Enchantment, which bears no resemblance to the tensions of daylight and alarm clocks, dentist appointments and difficult clients.

# Chapter 19
# The Overlook

Where the creek passes under the Drive it has always beckoned the passerby to slide down the earth embankment and stand on top of the culvert to gaze into the water, or out across the floodplain. It was fun to sit on the bank with my feet on the top of the culvert listening to the gurgle of the current and watching the variety of birds, each doing their thing in and out of the cedars and dogwood along the creek. For several years I watched the embankment erode back to where the top was only a half a foot from the pavement.

It was only a matter of time before the county highway department would notice the erosion problem and set out to fix it. I decided it would be better to bring it to their attention and learn what they proposed to do to stop the progressive erosion and provide a permanent fix. I didn't want to be faced with a fait accompli which I didn't like and would have to look at forever. A visit to the highway commissioner proved my concern was well-founded. When I asked if they might build a bridge abutment, the commissioner's reply was all they would do was place riprap on the embankment, "No money for anything more than that." Whatever natural aesthetics the embankment had with its vegetation would be lost forever under the rock.

By this point the reader is probably anticipating that I was thinking about building something of my own design. And I had been. A year or two prior it had crossed my mind that it would be nice to have a stone headwall at the level of the pavement, which would overlook the meandering creek and its floodplain. From the creek the wall would appear like the side of a stone bridge. When I casually mentioned this to Jan, she didn't think much of the idea. When I again mentioned the subject to her after having seen the highway commissioner, she realized the seriousness of my intention and asked me rhetorically if I'd lost my marbles.

With this encouragement, I proceeded with a sketch of what I envisioned to be the appearance of the overlook; from the road a

27-inch-high weatheredge limestone wall 16 inches in thickness, with a natural limestone cap. From the creekside it was to be a typical stone bridge sidewall with a recessed arched opening surrounding the culvert. The highway commissioner met me at the site to look over the sketches. He said he thought it looked great (whatever it looked like, it had to be better than riprap). If I wanted to do it it was OK with him, but he emphasized that the county would not pay for it.

This is all it took to get me going on the structural design and the details of the architectural appearance. Before I did this, I called Rick, the contractor who had done the beach protection work. The all-important question was how do you do all this work beneath the water level and surrounding the culvert without stopping the flow of the creek. Damming it, even for a couple of days, would result in flooding upstream properties, including some of my own.

The complexity of the engineering problem was a great part of the fun. My idea was to build a cofferdam out of steel sheet piling driven to bedrock only five feet beneath the surface of the creek. The cofferdam would be 22 feet long (the length of the wall) and six feet wide with its back side up against the pavement and its front just beyond the mouth of the culvert. It was Rick's idea to strap a gasketed extension to the end of the culvert to carry the water into the creek beyond the cofferdam during construction.

When all the earth was excavated from within the cofferdam, the inside was lined with plywood and a concrete foundation poured to just above water level. It was strange looking down inside the cofferdam watching people work several feet beneath the water level just the other side of the one-half-inch-thick steel. If left without pumping, the "hole" filled up with water in a matter of several hours.

This was a very high-visibility project. It closed off the northbound lane on the Drive. The crane and excavation equipment and materials were necessarily staged along the side of the road. After the foundation was poured, Rick was to pull the sheet piling out and form and pour the concrete retaining wall up to the grade of the road. The area between the pavement and the wall was to be backfilled, and then the equipment could leave the job. The masons who did the fireplace and chimney on the house would subsequently face the concrete with stonework from the waterline up to the top of the retaining wall, finally laying up the full stone wall above the grade of the roadway.

All this was scheduled carefully with Rick and Yukon to ensure completion prior to the onset of cold weather, which would preclude any masonry work for several months.

More often than not when building things, the best organized plans go haywire. And this was no exception. I received a call from Rick that the DNR's local representative had stopped the work and arrested him for "dredging" without a permit. A small amount of soil excavated from the cofferdam had fallen into the creek, and awareness of what was going on reached the DNR. Rick should have known the project necessitated a permit from the state, even though the county highway department had given us its blessing to undertake this construction on its right-of-way.

I encouraged Rick to get the fine paid quickly. It couldn't be paid for several days because the bureaucratic paperwork had to catch up with the citation. To get things back on track I suggested that he walk the permit through the approval process at the DNR regional headquarters in Green Bay, because their normal permitting time is two to three weeks.

I called the DNR in Green Bay and spoke with the appropriate official (whom I had previously met when the shoreline protection work was underway). I apologized for the lack of a permit, which was an oversight (at least on my part), and said that I would have the soil spillage cleaned out of the creek. I explained that the last thing I wanted was a problem with them when I was trying to do something to enhance the scenic vista along the creek and the road, besides eliminating the possibility of any bank erosion (which doesn't do the water quality any good). I reminded him of all my past work on the property, and that each time I had consulted their department on procedures and permits. I jogged his recollection of the swamp restoration and other measures I had taken to protect the environment, which many people would consider extreme. Rick had given him a copy of the engineering drawing I had prepared, so he was aware that this wasn't a slap-dash solution.

I also asked if Rick could walk the permit through because of the advancing cold weather problems we would be facing with masonry work. The official said yes, but it would take a couple of days, at least.

A couple of days turned out to be three weeks of stonewalled frustration. I made several calls and always got friendly and polite responses; like, "The typist was on a short vacation," or, "It's between offices," etc., with assurances it would only be a day or so. To Rick's inquiries the replies were barely civil, if not openly hostile, to the point he did not want to call again. His reluctance to aggressively pursue the permit stemmed from an entirely different perspective—one of having to take out future permits from the same source.

It didn't matter that there was an open hole 22 by 6 by 9 feet deep along the side of the pavement, nor that the roadway was partially blocked, nor if below freezing weather did arrive early the project might linger on until spring. The regulatory process had been flouted, and the bureaucrat's revenge was the labored issuance of the simple permit.

God knows why neither Rick, nor I, thought of the need for the DNR permit, but as the DNR official said, "Rick should have known better." You could say the same about the county highway department. And when you get right down to the nub, I'd certainly been involved with a lot of permit processes for twenty-five years and should have anticipated this.

So the work lay dormant for four weeks while a beautiful autumn faded away. People asked why nothing was being done so I owned up to the permit imbroglio and said lamely that I thought it would be all resolved "in a few days." I'll always wonder if it was a neighbor who blew the whistle to the DNR, or if they just happened along, not that it matters.

All the while Yukon was chomping at the bit to get his masons on the job before freeze-up. When the structural work was finally completed and Rick's men were gone, we still had to wait a week for the masons. They were busy completing another job. Don got everything ready, building a scaffold across the creek next to the headwall. The stone had been piled there for several weeks and was beginning to walk away. Fortunately, the pieces were too big and heavy for the casual thief, but one night as Muggs and I came outdoors, I did hear the scrape of stone against metal. Before we could get out to the Drive, a pickup loaded with my stone was pulling away.

Late November's usual stormy weather disappeared in a fortuitous interlude of unseasonably warm and sunny days. Ideal weather to be hefting around and setting stone. As the entire structure was to be given its aesthetic character by the masons, the consummate skill and artistry of the stonework was ultimately important.

The landscapers dry-laid large weatheredge limestone along the creek banks either side of the headwall until they reached undisturbed soil. On top of this stabilized creek bank they placed field sod up the embankments to the roadside. This heavy and thick (six inches) sod had been dug from a pasture at Greg's and contained the most wonderful variety of weeds and plants. By midsummer the following year, it looked as though it all had been there forever. The scene was framed with five large cedars, spruce and a white pine, transplanted to the shoulders of the road flanking the wall. This helped limit the visual focus downstream and out over the floodplain to the view from the overlook. The entire area of plantings and sod was surrounded with snow fencing until early the following July to prevent the springtime fishermen from tearing up the sodded slopes.

It occurred to me months after the fact, that I had the ingredients of a legitimate gift to the county, which could be written off from that year's income. The county highway commissioner had the highway committee come by on one of their tours and had a most favorable reaction from its members. The letter of thanks from the county was combined with copies of all the bills I had paid and forwarded to the IRS, resulting in a very beneficial refund... for once.

Quite a few of the neighboring people along the Drive made an effort to let me know how often they paused on their walks or rides to lean against the wall or sit on top to enjoy the view of the creek. Though I dreamed it up and had it built as much for my enjoyment as for others, these accolades were nice to hear–to be made aware that others found it pleasing and fun.

I do so enjoy sitting on top of the wall with Muggs beside me, listening to the gurgle and splash of the water and observing the "goings-on" of nature along the creek and the floodplain. Looking at the face of the headwall from the creek or woods, the stonework and its structure comprise a complementary background to the variety of green, growing things beside the creek. In the colder times of the year, my contemplative sits on the wall are abbreviated by the stony chill penetrating through the seat of my pants. Yet, it's worth it, because the peace and contentment remain within me far longer than does the cold in my rear end.

The short and somber days of early winter bring with them a stark palette of color along the creek. The soft whites of early snow cover are bisected by the creek's inky black surface and are contained by the slate-like gray of the lake beyond. The scene is compressed under a gloomy monochrome sky, so low and heavy that I unconsciously hunker down into a stooped posture to pass clear under the leaden overhead... and to keep the raw wind from knifing down the collar of my jacket.

Springtime gives birth to bright yellow swamp buttercups growing along the banks and the congregation of the spawning fish in the deep pool just beneath the overlook. The rapid growth of tall grasses and weeds in summer nearly choked out the newly planted tamaracks, until we cut around them. These same weeds give the floodplain a delightful variety of speckled color, and nourishment for wildlife. I have seen great blue herons wading elegantly in the creek, and one evening came upon two does up to their tummies in tall grasses, feeding nonchalantly along the floodplain. Once accustomed to my presence sitting on the wall, silent and still, they decided I wasn't a threat and went back to their meal. Only the mosquitos drove me away from my seat, sharing their world.

# Chapter 20
# The Garage

It had always been my intent, however vague, to someday build a garage. I had even shown the future garage on the original site grading plan. Its location was depicted on the same side of the driveway as the house, only about 80 feet distant. As illogical as it may seem toward the end of the twentieth century, an attached garage was never within my scheme of things when designing the house. I didn't like the idea aesthetically and, from a practical standpoint, any greater building length would have meant losing trees, which was unthinkable.

Though accustomed to the convenience of a garage (an attached one at that), I didn't mind the outdoor parking at the Schloss, even in winter. There were times, of course, when the temperature was far below zero, and we wondered if the car would start in the morning– but it did. Brushing the snow off was tedious only when having done it first thing in the morning; by afternoon it had become a repeat drill.

After a few years it was becoming more difficult to find places in the house to store tools, kindling and garden hoses, to mention a few items we had accumulated. Hoses, paint, shovels, rakes, ladders and reinforcing rods for guiding the snowplow along the driveway all spent their off-seasons in the crawl space. Down in the crawl space, shoe-horning these things in and out of the hatch, I knew exactly the Culligan man's feelings, and I was down there far more frequently. Getting the stepladder in and out was always accompanied by bruised knuckles. There were things that simply would not fit through the hatch. The wheelbarrow was covered with Visqueen, secured with duct tape and left upended behind the wood piles. Outdoor furniture, kindling wood, skis, and an eight-foot artificial Christmas tree were stowed in the attic. We really needed a garage!

Being aware that a covenant in the deed states that only attached garages are permitted, I was somewhat put off about the whole thing. Of course, another covenant stated that only people of the Caucasian race are permitted. It's an old deed.

Both restrictions, obviously, were intended to maintain a superior standard of living along the Drive. The former by preventing a proliferation of unsightly shacks on these up-scale residential properties, the latter to keep out the riffraff. The latter deserves no further comment. But, the irony of this well-intentioned architectural restriction was, though it prevented for the most part cheap and ugly outbuildings, it forced people into building ugly houses, driven by the singular requirement for attached garages. For instance, a common design has been to place the garage on the first floor and the living quarters above. There are all sorts of variations on this theme. A few homes had been tastefully done where differences in grade created the topography suitable for a multilevel architectural treatment; the rest being compelled solutions to accommodate the garage.

Despite the covenant, there were more than a few detached garages–twenty-one by count. There may have been more that were totally hidden from view. Obviously, the covenant had not been enforced. Enforcement takes time and money. As I got more serious about the garage, I decided that a visit to the county zoning department was in order, because contrary to the covenant, the ordinance permitted detached accessory buildings in this zoning district. One unusual paragraph in the zoning ordinance limited the height of detached accessory buildings to no more than 14 feet. This restriction would prevent me from designing a garage compatible with the architecture of the house, an absolute necessity from my viewpoint. The garage was going to look as good as its principal building.

I had decided I wanted a garage with only one auto bay, but it had to have a large area for storage. As the roof pitch on the house is 12:12, this was going to dictate a garage ridge of 18 feet, well beyond the permissible height. With a plan and elevation drawings as convincing evidence of the architectural worth of what I was intending, my visit with the zoning administrator was informative, but discouraging. I might as well have brought in a sketch done by a third grader on foolscap.

I learned several things, only one of which would help me in my quest to build a proper garage. The positive note was the county did not enforce private covenants. The negatives were at the time insurmountable. The height limitation was the result of a misguided attempt to prevent people from converting their garage attics to rooms which could be rented to tourists. It had been a piece of poorly thought through, reactive legislation to a real problem. Its consequence was to prevent anyone who owned a house with a steeply pitched roof from building any detached structure compatible with the architecture of the principal building—which would make all the sense in the world for enhancing the appearance and value of the improved property. If anything, there ought to have been an ordinance dictating the match of accessory buildings to the architecture of the main structure. But the logic of that would have been overridden by the common wisdom that it would be infringing on people's rights to build what they want, architecturally compatible or not.

Neither of the avenues available to overcome this ordinance impediment seemed very promising. The zoning administrator felt that seeking a variation to the maximum height would not have much chance of being approved by the appeal's board. The alternative approach would have been to propose a change to the ordinance. I felt that as a nonpermanent resident with an individual axe to grind, I would have little success with this approach. It had to be done by a local person, with a background of experience in this sort of thing, to favorably influence the county board.

The zoning administrator mentioned a local architect who had voiced a similar objection to this awful ordinance. There's my guy. A phone call brought an enthusiastic response. He also confirmed that all the other architects in the county felt the same.

To stimulate his interest in doing something, I said I would pay for his time. Realistically, without a change in the ordinance, I could not hope to have a compatible detached garage. I felt strongly if it could not be done right I'd forget it.

So my garage planning ground to a halt awaiting a response from my architect. After nine months and numerous phone calls, I ran far beyond whatever patience I was blessed with and simply gave up. Until one day I read in the local paper that an attorney I was acquainted with had been elected to the county board. A visit to his office with my drawings, and an explanation of the negative implications of the present ordinance to the design of buildings county-wide, generated the preparation of a replacement ordinance by the county zoning staff. It was deliberated on and passed into law within two months. The new ordinance recognized the desirability of compatible architecture. Conversely, it addressed the need to prevent unsuitable tourist housing by the sensible and straightforward prohibition of plumbing in accessory buildings and windows in any second floor or attic area of these buildings to preclude habitability.

For me, this was the easy part of the equation to have resolved. The more difficult aspect was violating the covenant. I deliberated equivocally with myself on this for a year or so. I belonged to a property owner's association whose goal was to watchdog what was going on along the Drive. It monitors governmental policies and activities affecting the property owners' interests, and uses the clout afforded by the aggregate membership to favorably influence our interests. I was acquainted with several of the association's officers and really didn't want to become an enemy in their midst, gleaning their enmity, if nothing more serious. The flip side of my indecisive deliberations was the fact that the covenant had been violated at least twenty-one times. So from that standpoint, I didn't feel so bad about building my garage, wherever. When completed, it would be a visual asset to the property, regardless of its location.

One of the neighbors who was aware of my predicament said, "Oh, just go and do it, the hell with the covenant." I guess I'm just not made up that way, despite that nearly a score of other property owners obviously had taken this attitude.

Another alternative I considered was to put the garage on the triangular piece of property the other side of the access lane. This solution didn't take long in coming unglued. It skirted the covenant, because it was on a separate (yet contiguous) tract. But, for the purpose of development, the county zoning ordinance defined contiguous property on either side of a public right-of-way as separate properties. Accessory-use buildings are not permitted on properties without a principal structure. Gottcha! The creek meandering through the triangular property posed another regulatory problem as another ordinance says you can't build within 75 feet of a navigable stream. By the DNR's preposterous definition, this thread of water through the forest met the criteria. I asked the DNR representative to confirm this on-site, which he did. The only high (dry) ground suitable for building necessitated a ten-foot variation from this ordinance. I went so far as to obtain a topographic survey and prepare a site plan to substantiate the reason for the variation. While preparing the documentation for the variation I had an antipodal change of mind. This was ridiculous going through all this to put the garage an impractical 200 feet from the house! So I scrapped this idea forthwith.

I quickly had a pencil rendering done of the garage to demonstrate the architectural thought I had put into it. It also illustrated its harmony with the house and its surroundings in the forest along the driveway. I took my rendering and visited with the president of the association telling her what I intended to do, and why. She was very nice, although a bit flustered as she realized the association was being put on the spot. I left the rendering for her to show the other board members at their next meeting. Subsequently, she called and said the association would not formally object to the detached building.

In retrospect maybe it would have been better to just say, "the hell with it," as my neighbor suggested. The association's decision concerning the garage was astute in the overview of homeowner's interests along the Drive. Should they have fought it, neither of us was in a terribly strong position. The validity of any objection would have been weakened by the multiplicity of precedent. A fight could have opened up the proverbial can of worms with others whose improved properties violated the covenant; to say nothing of it being costly to pursue in a legal milieu. Should I, in the end, have been successful, it would have effectively broken the covenant, with future development implications far greater than created by my singular garage hidden in the woods.

The resolution of this potential conflict within myself gave impetus to finishing the construction drawings. To the layman, the effort involved in preparing the architectural drawings for a garage may seem cursory. Most people build garages without detailed drawings. But not this garage. To keep it consonant with the character of the house and adapt the architectural theme to the intended uses, four sheets of details were necessary.

My workbench in the utility room, though ample for use as such, has remained as pristine as the day its top was painted. Flower arranging and shoe shining have been the most destructive activities engaged in at this "work station," which is too close to the entry and too visible for Jan to tolerate any uses that would have resulted in the top becoming beaten up in typical workbench fashion.

The garage was to provide me with an honest-to-God workbench. But I didn't want to crowd the spacious auto bay. The solution was a box bay into which the bench top was partly recessed. In effect, the top was designed as an unusually deep window sill. The window exposure to the out-of-doors was on the north side so the triple sash would flood a wonderful, even light across the surface of the bench. Needing a wood vise and a metal vise, I asked Mom for those that were part of my father's workbench, unused for many years. These pieces of equipment, much older than I am, had serviced me well in the basement of my parent's home when growing up; making all sorts of things from ship models to molds for counterfeiting nickels out of lead. It is with some nostalgia that I again look upon and use this equipment at the Schloss. I think it would make Pa, the maker of many things, happy to see his tools in use again.

Jim was most relieved when he learned I was actually going to get underway with the garage. He had been storing the framing lumber since the tornado, when the pines from which the lumber originated had been cut up and milled in his cousin's sawmill. Beside the 16-foot-wide auto bay under the central ridge, there was to be an adjunctive shed roof extension to the building on the side opposite the box bay. This space would give an additional five feet of storage depth inside and firewood storage under roof on the outside. The inside storage bay was to be totally open to the main part of the building, the roof held up by white pine beams and columns milled from the trees lost in the tornado. The shed roof was recessed from the front to make space for a man door into the main part of the garage. Both this door and the overhead door were clad on their exteriors with vertical cedar boards nailed to the "real" door with old-fashioned cut nails in a precise pattern. Each have large iron strap hinges painted black. The man door closes with an old-fashioned thumb latch. The overhead door has a cedar doorstop down the center, flanked by two, large black iron rings, as though it opened inward like an old barn door, rather than upward, raised by the convenience of an electric door operator.

The crown on this little shingle-sided, white-trimmed building was to be a white cupola, the details of which occupied one full drawing sheet. I wanted Jim to do this in his shop for Don's carpenters to install. It was a fussy thing to build, and large enough to be much too heavy to lift into place on the peak of this steeply pitched roof without a crane. Because of this, I designed it in four sections. The lower base was to be attached to the roof and flashed over with copper. The upper part of the base was to fit snuggly down on top of the flashed base.

Likewise, the box-like louvered section was to be set on top of the upper base. Overall was the pyramidal roof with projected eaves and cedar shingle roofing. The entire assembly was held together and kept in place by a long iron rod with a threaded end bolted to a collar beam in the attic. This assembly was also necessary should the cupola ever have to be removed due to storm damage, rot or whatever. The cupola roof was to be topped by a large, copper weathervane with a single-masted sailing vessel as its monogram. But wait, why not remodel the sloop into a yawl (as is *Aurora*) to give it real meaning, if only for myself, the designer. So the off-the-shelf weathervane sloop went to PJ's to have its rig converted into a two-masted vessel. In so doing, they discovered the manufacturer's welds were so weak it would never have withstood winds of winter storms. As it was, the first winter a dead branch, broken from a tree during a storm, tore the jib off, so I had to get up on the ridge of the cedar-shake roof to lift off the weathervane. It was a stretch, barely reaching the weathervane atop the cupola. The whole rig went back to PJ's for further strengthening. I had been terribly ambivalent about the cupola as part of this building until I saw it in place. My first glimpse erased any doubts as to its use as an element in the design.

 The garage didn't end up where I had originally envisioned it. After I had staked out its location, the zoning officer said one corner was approximately eight feet too close to the setback requirement of the access lane. Sure enough, the architect had made a mistake. Unfortunately, the building location could not be shifted because of the proximity of large trees. I had never considered the other side of the driveway as an option, as it was much more heavily wooded and sloped. But, I was desperate for a solution which did not involve cutting trees down. Despite this constraint, I found a place for it on the other side of the driveway. I am a firm believer that acceptable solutions are almost always present. You just need the persistence to ferret them out, along with some creativity to envision them in situ, so to speak. I discovered a spot in the forest without any significant trees (and my definition of significant was rather extreme). It was just large enough for the footprint of the garage with the overhead door fronting on the center of an opening between two large overhanging cedars. The change in grade offered an opportunity for a low, stone retaining wall similar to that behind the house. Between the stone wall and the wood storage alcove could be a flagstone walk for wheelbarrow access to the firewood stacks.

Since the foundation was to be a monolithic, reinforced structural slab, I eliminated the need for the deep trenching of foundation walls and footings, which would have damaged a host of roots from the surrounding trees. The well-drained, sandy soil alleviated any concern for the possibility of frost heave without foundations below frost depth. I retapped some of my old sources for windows, copper work, decorative iron and exterior lighting fixtures to complete the architectural match.

The convenience of controlling the outside lights on the garage from the house, coupled with providing for future heat, dictated multiple wire runs from the house. Exiting the house with future electrical had been thought of when it was built, but that's where anticipatory engineering had stopped. The only way to get between the two buildings with buried wire runs, not slicing off zillions of tree roots, was to exit the garage under the middle of the overhead door and then ditch up the center of the winding driveway. The 125-foot length of the cable run required hefty wire. The power needs and three-way switching required five runs. Roger ordered 750 feet of wire from a supplier in Green Bay. After seven weeks of waiting, the garage was going to be finished before we had the wire. We weren't getting anywhere with the supplier, despite many promises.

"Tell me what to get and I'll get it in Chicago in a day, and bring it up with me."

This time it didn't turn out to be quite that easy. It took phone calls to twelve suppliers before I located one with sufficient wire in stock of the type I wanted. I dashed to the warehouse on a Friday afternoon to beat closing time, only to learn that I had been misinformed. It turned out to be aluminum wire, rather than copper. Returning home through forty miles of rush-hour-traffic I was a very unhappy guy. Score one for Roger.

Another four phone calls with even more explicit specifications netted the wire at a manufacturer's distribution center on the following Tuesday after our return from the Schloss.

The construction took far too long. Don had promised two carpenters, but that only lasted until the heavy-duty work was over. When the trim work, which was by far the most time-consuming element of the carpentry, got underway two on the job was spotty. Not enough was accomplished in between our visits. It was an easy thing to pluck a man (or two) off this job, when scheduling on other jobs demanded additional carpenters. However, the tedious progress was not sufficient to spoil the joy of watching the structure materialize. Having started the framing in June (two weeks later than promised), by September the unfinished job had gotten to be an irritant, and I let Don know. The squeaky wheel gets the grease, it is said. By the end of September, the shake roof and shingle-sided exterior were finished, the trim was completely painted, the barn lantern and copper gutters were installed by Roger, and the protective shrouds removed from the trees by Greg.

The interior is completely sheathed with a rough grade of 3/4-inch plywood and painted white. The plywood sheathing covers fully insulated walls and ceiling. The painted surface looks barn-like and makes the interior bright and airy. I can hang anything anyplace on the plywood... tools, cabinetry or whatever. The attic is floored over and is reached by a drop-down stair—a roomy storage space, avoiding clutter down below.

Some guys never read the plans. Roger installed the copper elbow spout from the gutter over the entry door incorrectly. He volunteered to redo it. Wrong again. This time I left a large sketch tacked to the wall, but they only got it half right. When this fifteen-minute job of braising copper was botched a third time, I gave up. It worked, and the fact that it wasn't pointed in a 45-degree angle from the building ceased to be important. Only I would know. I must have been mellowing.

Garages are by nature utilitarian structures. I had not wanted this one to fall in the prosaic category, and I haven't been disappointed. I sometimes park myself on the chopping block nearby and study all the details that give this little building, nestled under the boughs of the guardian trees, its distinctive character. The sentient pleasure from merely a glimpse never dims.

# Chapter 21
# The Access Lane

Turning into the driveway from the access lane on a Friday night in April, something on the lane, momentarily in the headlight beam, did not look right. I made a mental note to check on it in the morning and promptly forgot until the next afternoon, when Greg and I were walking down the lane discussing the program of spring work. It was instantly apparent that someone had been trimming trees along the sides of the pavement. The periphery of the winding gravel lane had been shorn of vegetation. Though the pavement was not straightened, the view had been.

Within a few steps we came upon a large cedar planted only two years prior. It had been stripped of all its branches on the side toward the roadway as far up as a chain saw could reach, mutilating a beautiful tree. On the other side of the lane, a cedar had been cut down. With this discovery I was becoming apoplectic and afraid of what I would find further along. And my worst fear was realized. Two birch trees (one more than six feet off the gravel) had been cut down, and on the other side an ash and a maple, neither of which were closer than seven feet to the gravel. This continued along the entire width of the lane, a quarter mile. It just happened to be worse adjacent to my property as this is the windiest, woodsiest stretch.

Coincidentally, a neighbor had recently purchased over 400 feet of property flanking the access lane on both sides somewhat further north. The trees on his property had been terribly ravished by the tornado, and the purchaser had his bulldozers (he is an excavator) cleaning up the mess of upended root wheels and thicket of tree trunks and branches, lying like an obstacle course across the property.

I called to ask if they had done the number on the access lane right-of-way. "No, but the county had a crew working along there for a couple of days," was the reply.

In the next breath, my neighbor's wife told me the county was going to widen the pavement!

# The Access Lane

It was hard to believe the county highway department, on its own initiative, had done this, and further, intended to widen (read straighten) this seldom-used lane. I had had more than one conversation with the highway commissioner concerning the snowplowing on the lane. He had been cooperative in keeping the plowing path narrow to prevent damage to the flanking cedars and spruce, after that first frustrating year the county had plowed. Yet, it had all the earmarks of a county job, given their mindset. I called the other property owners to let them know what was going on. One was in California, but said he would do what he could. I was sick about this the entire weekend, and exhausted myself trying to determine a way to stop this ridiculous road improvement.

At 7:30 Monday morning I was at the county highway commissioner's office. He has always been a polite person, but I wonder if it's only my imagination that senses he wishes he had an escape when he sees me coming. After I related the incident, he said he was unaware of any work being done on the lane by his crews. But added, "Wait a minute, I'll call my superintendent on the radio."

"Are you doing any work out on that access lane?"

"Yeah, we're all done with the cutting," the voice on the radio replied.

"Who asked you to do the work?"

"The town chairman."

"Thanks, 10-4." The commissioner looked at me with a weak smile and shrugged his shoulders.

I asked where the chairman lived and barreled out there, the speed of the car rising with my rage.

I got lucky. He was still at home. His wife let me in and we chatted, as the chairman was on the phone. Only a counter was between us, and his end of the phone conversation overrode our passing the time of day. Putting two and two together I quickly perceived he was talking to my neighbor in California and that the conversation was not making his day. When he hung up, instead of talking to me, he hurriedly made another call–to the highway commissioner.

"Did your guys do the work I asked them to do?"

Silence while he listened to the commissioner.

"Oh hell, I've got the property owners all yelling at me." He did not know the next one was sitting in his living room.

After several other comments, the phone conversation ended something like this. "Yeah, those guys out on the Drive are all a bunch of jerks. They like trees better than people."

That people and trees are incompatible was a revelation to me.

This man's day was off to a bad start, and I was soon to aggravate his disposition further. I made a conscious effort to start on a positive note, which lasted all too briefly. I suggested that the property owners would have liked to have had some input, or at least had been made aware that improvements were being considered, prior to the work being done. We then could have told him that the aesthetics of the lane were more important to us than having an improved roadway, which is an invitation for others to use. His reply was terse at best, "I'll maintain the town roads as I see fit!"

Whoa—and he was standing for reelection in a week or so. Shows you how much clout the thirty registered voters along the Drive have.

I explained I had gone to considerable effort and expense planting trees to beautify the right-of-way, some of which the county crew had mutilated or cut down. It was spoken to deaf ears. He intimated that the snowplowing was more important. I made him aware of my snowplowing "agreement" with the highway commissioner. He clearly registered disbelief. By this time his wife looked like she wanted to crawl in a hole to get away from this acrimonious debate going nowhere. The chairman's face was becoming florid, and there was no holding back his frustration. "We'll give the road back to you property owners, then you can maintain it. It's just a big pain in the ass." (I know who he thought was the "pain in the ass.")

"OK, you'll hear from us," and I exited the debate and the door as gracefully as I could.

He didn't know, but his wrathful comments had triggered into action, from wishful thinking, my long-held desire to get rid of the town's right-of-way. But, would the other four property owners feel the same? That afternoon I saw an attorney to ascertain if such a thing could legally be accomplished and what the procedure would be. It was, and the steps were well-defined, but without the support of the other property owners along the right-of-way, this initiative would revert to nothing but a pipe dream.

The attorney made me aware of another impediment; the annual grant of funds towns receive from the state for road maintenance.

This short road had to be a moneymaker for the town as it went years without any maintenance except the occasional plowing. The county did the plowing for the town, charging considerably less than a private plower would. For a quarter-mile road the grant money was not a big-time consideration—just a potential excuse.

I hadn't a clue why privatization of the access lane would be advantageous for the other four property owners. None seemed to give a damn what went on, except for Gene. In fact, retrospectively, I cannot help but make a suppositional connection between the timing of my neighbor's clearing of the tornado-damaged property and the coincidental roadwork orchestrated by the town chairman. He just didn't seem to be the type to have undertaken this on his own initiative. And, despite promises to the contrary, this neighbor never did cooperate on calling the chairman to help give weight to our complaints.

Possibly Gene would have some innovative insight into this conundrum when, in a few days, spring would be bringing him home from California. He did, and his reasoning made sense for each of us. Should the lane remain as public domain, the property owners would have nothing to say about its maintenance. Though we might get this town administration on our wavelength, another chairman would have his own ideas on maintenance, widening or paving. As it is, the properties along the access lane have much more privacy than do those fronting on the Drive. Improving this lane could be perceived as an invitation for more intensive use as a shortcut, as the population and traffic along the Drive inevitably increase with time. Better that the destiny of the lane be in our hands, though it might cost a little money each year to maintain.

Gene volunteered to approach the other three property owners if I would handle the legal part. For one reason or another, everyone agreed with him and signed the petition to the town board to vacate its right-of-way. The law required an advertisement in a newspaper and notification of intent to other nearby property owners. Since property owners other than the five petitioners had no direct interest, this wasn't of great concern as an impediment. On the other hand, the chairman, if insincere, could be. Indication of this was clearly evident in informal encounters between our attorney and the chairman at various governmental functions. I suggested to the attorney that he put an end to these posturings by suggesting to the chairman if he wasn't going to support this, he ought to be candid about it, as the property owners were undertaking the necessary legal expenses based upon the chairman's stated desire to get rid of the road. The attorney did so, and it ended his bombastic threats of rejection.

You really want these kinds of things all wrapped up, if possible, prior to the formal hearing. This certainly wasn't the case. We were not at all sanguine about our chances of success right up to the vote. Other than the attorney, only Gene was present at the board meeting, which by his words was a nonevent. Without any preliminary discussion the board voted unanimously in favor of the petition.

Thus ended the first half of the attorney's work. The second was to draft an easement agreement to guarantee access for the interior property owners to their homes along the now private lane, and to establish some kind of minimal maintenance agreement based upon a pro rata cost sharing. The only mandatory maintenance included in the agreement was snowplowing. It took five months to get everyone to sign-on. It didn't matter to me as I was out at one end with property on both sides of the proposed easement. One of the other property owners suggested we get the county to plow the snow. Not in my lifetime!

I'd postponed all the work I had wanted to do along the access lane until the outcome of the right-of-way had been resolved, once and for all time. I certainly wasn't going to plant any trees along the shoulders. Nor did I want to restore the gravel roadway to its original narrower width, only to have it undone by the next pass of the county's road grader. After the right-of-way reverted to the property owners, Greg shoveled and raked the gravel up and placed cedar mulch along the verge.

In the late fall at our usual tree planting time, we planted an additional four cedars and white pines. This operation included dumping piles of sand at three of the four locations to raise the shoulder level with the road grade. Greg couldn't get into the fields with the tree spade truck to get the trees because of wet ground conditions, so the sandy piles lay exposed to view for several weeks.

During this delay I received a call from the Corps of Engineers investigating a complaint that "someone" was filling a wetland on my property. The Corps representative wouldn't tell me who it was that complained. It made me angry to think the complainer didn't have sufficient imagination to discern, or initiative to ask, what was going on before siccing the Corps on me. Among other things, I told the Corps representative I had been informed by the DNR that the area where the trees were being planted was not wetland. I suggested that he contact Greg to verify what I was telling him prior to taking a look for himself. A week later I got a call from him to tell me they had an obligation to investigate all complaints, but having done so he was sorry that it had caused me a problem. I replied that my obvious frustration stemmed from my awareness of several instances of gross abuse of wetland's regulations along the Drive. Each of these "improvements" was readily visible, yet wetland's laws had not been enforced. In this instance I was intending an incredibly inconsequential modification to the landscape, one which ironically was being undertaken to beautify the roadside.

It reminded me of the summer I had Greg place some rock (twenty or thirty rocks to be more precise) along the side of the creek where the current was eroding a steep bank upon which grew several small trees. My immediate neighbors saw fit to complain to the DNR, which directed me to remove the rocks posthaste or suffer the dire consequences delineated in their letter. These are the same neighbors who spend their summers digging channels and erecting ineffectual log barriers along the creek to prevent the occasional natural redirection of the stream across their beachfront.

A year after the access lane right-of-way had reverted to the property owners, I noticed on a Friday's arrival that the lane had been graded out, widening the gravel surface. I had always assumed the town board had notified the county highway department of the privatization of the road, so the county would not do any work for which the town would be obligated to pay. Will I never learn?

Assuming that someone has used initiative, followed through or acted responsibly is at best disappointment awaiting opportunity. Usually I just ask!

Unfortunately, I didn't in this case until the damage was done. My suspicion that no one had let the county highway department know proved correct. The commissioner said he hadn't been informed of the change, but was certainly glad to know. I suppose I could have demanded that the county restore the grassy verge it had graded over, but I didn't. My goal having been achieved, I was weary of the whole controversy. So I raked it back myself and had Greg place some additional mulch along the wayside.

# Chapter 22
# Holding Tanks

I was just beginning to relax from the rigors of winter, rapidly and comfortably accommodating myself to the tantalizing promises of spring. I had planned to concentrate on writing over the weekend and spend Monday morning at Palmer Johnson getting *Aurora* ready for its eminent launching.

Upon Friday evening's arrival I noticed the holding tank light in the utility room was glimmering its red warning—an indication to get the tank pumped out. The period of use in between pumpings is one of those things that you get used to with a subliminal anticipation of when it is getting full—you don't need the light. The light was indicating a condition which registered another warning within me that the tank should not have been so full so soon.

Returning from Muggs' early morning walk, I happened upon the pumper doing his thing. His usual friendly greeting was followed by, "I was going to come up to the house to see you."

"How come?"

"The tank's leaking. Come 'ere, I'll show you."

Peering down through the access hatch into the empty tank, two thumb-size spouts of water were clearly visible on the bottom. He maneuvered his large suction hose over the bubbling and dried out the bottom, only to have the leaks start again when the hose was removed. Sure enough, the eight-year-old steel tank had a porous bottom. That old sinking feeling welled up within me.

I needed a welder to go down into the tank, grind off the coating and patch over the holes with a steel plate dutchman. Disparing that there were any people around who would consent to undertake this odious task, I called the folks who installed the tank and was referred to a steel fabricator in Sturgeon Bay. He happened to be the successor to the man who made the original tank but had gone out of business in the interim. The new fabricator agreed to do the welding inside the tank at the rate of $100/hour and would be at the house by noon. In that it's not the most enjoyable thing to be doing on a delightful Saturday afternoon in spring, and that he was the only show in town, I swallowed hard and bought into the deal. He also needed a pumper to keep the tank dry of the groundwater welling up inside.

When he climbed down the ladder into the tank, he informed us there was another large leak spouting up five or six inches—the height of the spout indicating a lot of groundwater pressure. The whole region had just experienced two of the rainiest weeks imaginable. The nearby swamp was full of water. The flooding across the low ground was extensive, reaching to within 15 or 20 yards of the tank. What a time to have a leak in an inground tank! Between the time that the tank was pumped out at 8:30 a.m. and noon when the welder got there, this 2000-gallon vessel had filled three-quarters full. The groundwater was seeking its own level. Equilibrium was not much lower than the top of the tank, effectively rendering it useless as a storage container for all but a small amount of waste water from the house.

After two hours in the tank welding patches, he came out with a grim look to confirm my worst premonitions—he wasn't getting anywhere. He was creating more holes than he was patching. The tank bottom was pitted sieve-like from electrolytic corrosion.

It didn't make any sense to have someone creating holes in the tank at $100/hour. Before he came out I passed down some cedar kindling and a hatchet. He plugged up most of the holes which dramatically reduced the rate of inflow.

Besides being upset with the loss of the tank after so few years, I had a feeling the tank replacement was going to become a big deal. The tank itself was covered with a poured concrete saddle. The 18 inches of soil over the saddle would have to be removed, as well as the juniper bushes, which have thrived so well in this location. Greg suggested temporarily transplanting these into pots—large pots.

The saddle was to be broken up with an air hammer and its steel reinforcing burned off. If an unsaddled tank were emptied in a high-water table, it would probably pop out of the ground, buoyant like a balloon, with the groundwater and loose sand sloughing in from the sides to fill the void.

Subsequent to the tornado, we had planted three large trees in close proximity to the tank. Greg, the holding tank installer and I agreed that the excavation required to "unglue" the existing tank from the soil's suction, and subsequently, to set the new tank in place was going to cause a significant collapse of the surrounding saturated sandy soil. It is just like digging a hole on a beach; the more you remove the sand below the waterline the more it fills back in. Listening to this discussion, I began to have visions of these beautiful trees, now 30 to 35 feet high, with trunks 10 to 12 inches in diameter, tilting into the hole as the soil sunk under them. The holding tank installer remembered my concern over the trees voiced eight years prior. It was becoming clear to me now that he was uncertain of his capability to do the job without the high probability of doing significant damage to the trees.

I remained awake much of that night pondering my options. A concrete tank would be less apt to deteriorate in the soil. Yet, to get the same capacity, its overall dimensions would be greater than a steel tank, requiring a larger hole closer to the trees. Regardless of tank type, without shoring the sides of the excavation, the surrounding soil would collapse into the hole.

The only kind of shoring that would work would be steel sheet piling. Unfortunately bedrock was at the underside of the tank preventing the bottoms of the steel sheets from toeing into soil beneath the bottom of the excavation. The branches of the white pine and spruce were going to have to be carefully pulled up and held back to allow the pile driver rig to do its thing. With the crane and backhoes and dump trucks crunching over the driveway, I ought to be planning on a new one.

This was becoming a nightmare with no escape upon awakening. The very thing which I had sworn never to do again was going to occur—using the driveway for access of heavy equipment. It was back to protecting the flanking trees with 4 x 4s and snow fencing, except this time the activity was all going to happen on and just off the driveway's side.

By morning I had decided to "bite the bullet" with the sheet piling operation and its attendant cost. By 8 a.m. I had reached Rick, and he offered to come out and talk about solving the problem. So Sunday morning was spent figuring out a solution to this fussy job. They would create a steel cofferdam around the tank and, if necessary, anchor the bottoms of the steel with rods drilled into bedrock.

This would keep the surrounding soil in place, though pumping would be necessary to keep groundwater from filling the hole. After the tank work was completed, the steel sheeting would be burned off a few inches below the surface—the cofferdam to remain in place forever. So when (and note I'm not saying "if") there is a subsequent tank replacement, it can be done in a much simpler fashion.

As Rick's firm not only did sheet piling and marine work, but also sewer work and holding tanks, it didn't make sense to have the original holding tank installer involved. Dismissing him made me feel badly as he had responded so quickly to my predicament. (On the other hand, I wondered if the installation of the original tank had something to do with its abbreviated life.) At any rate, multiple contractors on a job results in carping and finger pointing with the divided responsibility and problems of coordination.

I raised the issue of permits, naively supposing that since this was a replacement-in-kind of an existing system which had failed, the permit process would be simplified or nonexistent. I was soon disabused of this wishful thinking. Not only was the permitting process for a replacement system as laborious as it was originally, but time had added steps with additional paperwork and formalities.

There was a holding tank agreement with the town which had to be signed by myself and notarized. This document had to go to the town chairman for his signature. A $250 bond had to be posted so that if you didn't empty your tank, the town would. I thought I ought to look back in my check register to 1984 to see if I had previously been extorted this deposit by the town. I had, but this town chairman had none of the records of his predecessor. Fortunately, I had the old check. Instead of calling it a bond or deposit (which the town holds forever, earning interest on it) they ought to simply call it a permit fee. The holding tank agreement then had to be recorded with the county register of deeds.

There was a second agreement to be executed and notarized; this to the county health commission, and a third document, which had to go to the state along with an engineering drawing for the state's blessing. I was informed that this normally takes several weeks. Deja vu!

"Can you hand carry it through?"

"Sure, but they're backed up a week on walk-throughs."

"Better get hopping, Rick."

The permit process was laden with new twists. One agency says another will accept copies of the documentation from the original installation. The recipient agency insists they must have originals. The first says they no longer have the originals. Consequently they have to attest to the authenticity of the copies. The state doesn't have any record of having received the original soil tests. The county has them, but they are nearly illegible copies from the designing engineer, not that the soil conditions of the millennia have changed in the past eight years.

When I called the county sanitarian's office to see if the necessary documents had been sent out, the clerk said, "I don't know why you are so concerned about the permit, yours is an emergency situation, and the work can proceed with the permit process following along."

"Did you ever tell the contractor this?"

"He never asked."

I decided I had better get a grip on my impatience and a clamp on my mouth.

Thank God I'd decided not to proceed at breakneck speed because this all would have driven me to distraction. As it was, we barely met the proposed start date anyway. The delay was for Rick's convenience, not mine. Maybe I'm just getting older (and smarter). "Mellowing out," my crew calls it.

Meanwhile, I hoped for some warm, dry weather to drop the water table in the swamp (hence in the tank). Even a few inches would mean a lot, but the days stayed cloudy, cold and rainy. I had to mentally meter the use of water. If the tank was pumped out on Saturday morning that became the time to do the laundry, wash my hair or whatever. If I didn't do these things, the tank was going to fill up on its own anyway. By Sunday morning it was back to rationed use.

The weekend before the work was to begin Rick called to let me know I had another problem.

"Now what?"

He had been looking over the swamp and discovered its surface was coated with what he thought was the telltale sheen of pollution, particularly that portion located close to my neighbor's mound-type septic system. This was more than 300 feet from the faulty holding tank, so it didn't seem logical that the pollution was from my tank.

Despite this, it was an unbelievable coincidence of bad timing, so far as my interests were concerned. No telling what twist the bureaucracies would come up with. To be fined for polluting would be a most galling irony.

My neighbor of the trenching and the encroachments, was going to have a stroke if the county condemned the septic system and forced replacement with a holding tank. I didn't want any additional antagonism, but the swamp becoming a cesspool was unthinkable. Having spent so much effort and cost to restore it and nurture the biota back to its former lushness, it would be unconscionable neglect to see it ruined. In that it is home and sustenance to birds and animals, its contamination would have been unbearable to me. Besides, it will sooner than later stink.

Indecision is the worst course. If summer dried up the swamp, the evidence so vivid on the water's surface would become far less discernable. It probably was going to become a finger-pointing contest anyway. But with the tank being replaced, I'd take my chances with the county. I called the sanitarian and asked him to take water samples from the swamp. Making sure to be at the Schloss when the sanitarian came out, we also informally inspected the tank work underway. With the elevation of the water in the nearby swamp, it had to be fairly conclusive to the sanitarian that any leaky tank would have filled, rather than emptied. I was so paranoid about this that I'd kept a diary of the entire disaster from the point of discovering the leak over four weeks past—witnesses, names, dates, events and remedial action. Donned in his rubber boots and gloves, the sanitarian filled the sample bottles and said he'd call me as soon as the test results came back from the lab in Madison. The age of the neighbor's septic mound didn't give me much cause for optimism.

Awaiting me upon my return to Illinois was a letter from an attorney, representing the town board, accusing me of blocking town roads and threatening dire consequences, if I didn't cease and desist. It felt like the "old one-two"—to the gut and then the jaw. Time out! I needed to catch my breath. By the time the sanitarian called giving the swamp water a clean bill of health, I'd calmed down and, as related in the next chapter, I had started the legal wheels in motion to quash a particularly bizarre accusation by the town government.

The work site at the holding tank looked like some strange

Holding Tanks

stageset under wraps. The flanking tree branches were trussed up with rope to the trunks or splayed back and held in place with twine to other trees, or to iron fence posts driven in the ground. Snow fencing cordoned off the driveway and the work area, and the higher branches of the trees were encased in canvas drawn around the foliage to eliminate or reduce any overhang snaggable by equipment using the driveway or driving sheet piling.

The lack of a sewer system and other commitments (sailing) kept me from being at the Schloss during the construction. So after my visit with the county sanitarian, which coincided with the first day of pile driving, I trusted Rick to watch over the activity and keep me informed, which he never does unless it's bad news. I hate getting his calls, "What's wrong, Rick?"

"You didn't tell me there were electrical lines buried under the driveway. We severed them."

I'd forgotten that the electrical power from the house to the garage had been trenched through the gravel. It never occurred to me that Rick's equipment would punch through the 12 inches of limestone base.

The inflection to the bad news was that it wasn't their fault; I'd never let them know about the presence of the wires. The obvious implication was that I'd be getting an extra charge. Rick was astounded that there were five separate wire runs.

Severing the wires necessitated a phone call to Barbara, the caretaker, as the contractor was now going to have to get into the bugged house. I suggested that Rick use Roger to do the repair work, as he did the original electrical installation. But reaching Roger is (if you recollect) frustrating at best. On the other hand, the amount of exertion a contractor uses in reaching a subcontractor, whom he was directed to use, is almost immeasurable on the scale of effort.

"Well, I tried him (once) and left a message, and he never called back."

Then you talk to the sub, "I called him back, but he wasn't in, and I left a message."

Since neither are ever in, this game of phone tag could go on forever, except that it goes nowhere... one attempt and give up.

It's phenomenal how fast and persistent they are in returning calls when the caller says (if applicable) if you don't get there to do your job, we are going to have it done by someone else and backcharge you. The response time is like lightning and, also is the charged nature of the conversation.

It's good I had also called Roger to let him know what was going on, and that I'd pay his bill. I finally got a ring after three-quarters of an hour of busy signals and, wonders will never cease, he was there.

The work progressed without any snags, and my optimistic nature felt this time everything might conclude without any undue problems. I'm ever the optimist. The evening the tank was removed and the cofferdam excavated I received one of those bad news calls from Rick.

"They made the wrong size tank and we can't get it in the hole."

It turned out the fabricator made the next standard size up from that which we had discussed. A 2100-gallon tank is not much larger than the original 2040, except that it was for some inscrutable reason two inches less in diameter. The volume being made up in a 24-inch greater length.

Visions of this taking at least another week longer flashed through my mind, as well as the question, "Why?" I had spoken to the tank fabricator at some length about the limitation in dimensions. I had prepared a drawing indicating these dimensions for Rick, and had encouraged him to order the tank the day I spoke to the fabricator. It turned out they didn't order the tank for three weeks. And when they did, they told the fabricator to make the one "you talked about with Mr. Geudtner." No purchase order, no letter nor drawing—just make it. Meanwhile, the fabricator evidently forgot the essentials of our conversation and did what he pleased, without any confirmatory inquiry. We not only had discussed size, but also the use of a thicker gauge of steel and cathodically protecting the tank from corrosion by the use of magnesium anodes. It's strange he remembered all of the things I had requested that were up-charges, a conveniently selective memory.

Rick and I discussed the options and decided that cutting the tank down in size to meet the permitted volume would be the quickest and least painful solution. Rick was at the fabricator's shop at 7 a.m. and called to say the modifications would be done that day, and the tank would be on-site by late afternoon.

What he didn't say was that the fabricator denied ever having talked to me. I must have divined all of this knowledge of tank size and characteristics. There was shared culpability, but Rick was in a position of no strength so he had to pay for the remodeling of the tank—another example of the vulnerability of verbal agreements. When something goes wrong, everyone suffers from memory lapse.

I was scheduled to arrive at the Schloss on Saturday evening, which left Friday to finish the project. They worked until the job was completed—at 7 p.m. When I arrived the site was clear of all equipment and clean of debris. Even Roger had completed his electrical work. The only evidence remaining of Rick's work was a bumpy driveway and patches of oil-soaked gravel.

Rick was looking for some money, so he and I walked over the site on Sunday. He agreed to have a bulldozer and two men at the house at 7:30 a.m. on Monday to blade off the driveway and remove the oil-soaked gravel.

On Sunday afternoon I got tired of looking at the trees all wrapped and trussed up. I got a jump on the landscape restoration work that Greg was to begin the following morning. I found only one branch broken, which was really to the credit of the construction crew. They worked with large, heavy construction equipment with only inches to spare. This difficult, fussy job was accomplished with a great deal of finesse.

The snow fence took a terrific beating, having been smashed into kindling in some areas. Better the expendable snow fencing than the trees.

Rick commented that the site protection work was extreme in measure, but admitted that it set the tone with a visual statement to those on the job of the seriousness of the customer's expectations. It was far more effective than simply admonishments of caution and care from their boss, they having become inured to these ongoing lectures.

With the tank purchase came a 15-year guarantee to be free of leaks. The only thing leaking was the guarantee. Being suspect and cynical of equipment guarantees in our business, I actually took the time to read the fine print. The escape clauses were classic. To retain the guarantee, the tank was to be cleaned and inspected annually by an authorized inspector and a written report filed with the manufacturer—

the forgetful fabricator. Over the course of fifteen years, these annual cleanings and inspections would probably cost as much as a new tank. But to get down to the slipperiest nitty-gritty, the magnesium anodes and their connections to the tank also were to be inspected yearly. A preposterous requirement as the anodes are attached to the tank bottom, buried under several feet of backfill. The stone backfill surrounding the tank is under the reinforced concrete saddle straddling the top. The guarantors intent was quite transparent—the "official" document was nothing but a caricature with not much humor. I only hoped the tank would have more integrity than its guarantee.

# Chapter 23
# Conservancy

An innate curiosity of where it goes and what's along the way drew me farther and further into the forest. Some say it's the largest roadless area in the eastern part of the state. Its 3100 acres and three lakes take awhile to become acquainted with and to find your way around. The first year we skied along a few trails, the track broken by others, our energy giving out before the track did. Since we had to backtrack out the same distance, it was a constant reminder of our limitations.

With twenty-four inches of snow on the first of December, the first winter provided incomparable skiing for a full three months. By the fifteenth of March the snow cover was getting spotty, still we persevered over ice, slush or bare spots. Temperatures around freezing make the snow stick like glue to the bottom of the skis, but higher temperatures don't. You just don't have quite the glide... not ideal, but still fun with just a sweater on. More often than not, we skied early in the morning or late in the afternoon on these longer, warmer days of late winter to miss the transition through the sticky, clingy freezing point. Little did we know that winter would be an anomaly. Never since have the snow conditions been anywhere near as consistently good.

For the two winters following, snow cover was almost nonexistent. These winters we walked along the frozen trails and across icebound lakes. We hadn't gone far enough to puzzle out the mosaic of trails, and I always wanted to persist farther than Jan was willing to go. My friend Tom came to visit one weekend, and I had Jan drop us off at a trailhead familiar from skiing. I was determined to walk beyond where we had ever skied and to find our way back to the Drive five or six miles from where we started. One of the property owners we'd met several times on the trail (he on his snowmobile, we on our skis) had given me a general idea of what to expect in the way of "now what?" forks in the trail along the way. Five hours and a lunch later, I began to have an intuition I had been along the path we were trekking and realized our goal was close. But it wasn't the end of the hike, because we had to stagger back down the Drive to get home. It was torture getting through dessert that evening, and on full stomachs, ten minutes of torpor in front of the fire was all we could manage. Give up! We slept as only one does when physically exhausted and full of fresh air. I think it's the only time I ever went to bed before Jan.

Our ever increasing explorations of the pathways through this wilderness kindled an awareness that we were living beside and enjoying a rare vestige of undisturbed forest and swamp. It is a "terra incognito" surrounded by the farms and resort communities of this popular vacation area. Little known, even in the county, except to one or two landowners and a handful of hunters who "tough it out" into the lakes to shoot waterfowl. Only a few of the residents along the Drive are interested enough to venture in from the periphery along the two or three most accessible trails. But, when paths become soggy and overgrown, anybody's pleasant outing becomes a "forced march" few are willing to endure. (Those are Jan's words—she's been on several.) It weeds out all but the most determined... me, not her.

One of the early Christmases we discovered a three-quarter-mile long trail through relatively undisturbed forest. At the end of this stretch the going became rather rough, through some bumpy land with more trees horizontal than standing; mostly birch, all victims of the tornado. The initial three-quarter mile was so pristine we returned several times during the week. At the beginning, a quarter mile in from the Drive, a For Sale sign was nailed to a tree. Out of curiosity, I called the realtor, discovering that the salesman I was talking to had done all the steel fabrication for me when the Schloss was under construction. Bill's description of the 110 acres was typically voluble and glowing. When pressed, he grudgingly admitted it was a bit pricy—outrageous was more like it! It was owned by a farmer who'd had it on the market for several years. With a realtor's enthusiasm, Bill let me know that a contract logger was about to begin to clear out all the "downed timber," like this was doing any buyer a favor. I was very skeptical and let my cynicism show. By mid-January the logging operation was in high gear, and by February it was becoming apparent the 110 acres was being logged off, cutting any and every marketable tree of all species. The transformation wrought in two and one-half months was cyclonic. What the tornado didn't destroy the cutting did, and it became too depressing to hike this way anymore. Bill called and asked if I was still interested in purchasing the land. I guess I unloaded on him that his client had destroyed the aesthetic appeal of the land, so it had nothing left for me. They wanted it both ways, the income from the timber and top dollar for the wasteland left.

This got me thinking about the overall preservation of these wildlands. Pitted against the developmental pressures, current and inevitable in this part of the county, the future of this wilderness looked rather bleak. There were over thirty separate landowners, some who, I was in time to learn, had simply inherited the properties and had never even made the effort to find their forty or twenty. There was probably an equal number of views as to its future.

It was at this point of time that the elephants of Africa were being poached relentlessly for their tusks. A siege against this noble species that was totally out of control in the quest to fill the insatiable Asian markets for ivory, which had skyrocketed in value to be worth more than gold. Thousands of these magnificent, highly intelligent animals were losing their lives each and every month, and the cry was just beginning to be heard. The environmental organizations were gathering their wits and their always inadequate funds to arrive at some effective strategy to stop this massacre. Publicity was just beginning to reach our consciousness, but not yet that of the world's policymakers. I, for one, cared so much it sometimes hurt. I felt so frustrated and helpless a half a world away. Donating money to those on the "front lines" didn't help assuage the pain inside from the ever increasing shock—it was becoming big-time news. To make it worse, it wasn't only the elephants of Africa. It was (and is) a repetitious dirge of environmental disasters and species loss occurring over the entire world. Though I couldn't make a detectable impact on the future of the elephants, maybe I could upon the future of this wildland outside my backdoor. A goal, if realizable, which would bring lifelong gratification and pleasure, not only to myself, but to others who appreciate and value the natural heritage of this small and picturesque place we call home. A place so popular it is in danger of being overwhelmed by those who love it and want a piece of it, and by those who want a piece of the action.

When perceived within reach, profit is a deadly curse upon folks whose motives might otherwise be more altruistic. I'm not pointing fingers or finding fault. In many instances, the sale of a piece of land may satisfy a real money need. Even if the heart is in the right place, the need for food on the table or school tuition dictates a purely economic choice. In these circumstances only someone who's lost his marbles would give away property with value.

I didn't even know if anyone else cared or how to go about this preservation campaign. One thing I was sure of was the only way this would ever succeed would be by purchasing land outright, not by becoming a spokesman for preservation, organizing an impossible coalition. But I'd never know unless I crossed the starting line. Take any tack, but start. One thing leads to another. After researching ownership in the plat book and at the county tax records office, I approached an owner of one hundred acres only to learn that it was being sold to the same property owner we had met along the trail.

If both he and I were interested in purchasing this land, our goals might be similar. Competing with one another could only drive value up. We spoke at length in his living room, hovered over by snarling trophies of big game conquests. He remarked that he wanted it all, but didn't think this goal was terribly realistic. He had a good start with several hundred acres. It was difficult discerning his philosophy except that he clearly didn't want to see the land developed. Certainly a measure of common ground, I thought.

It was a complete surprise when, a month later, I heard from this landowner, letting me know he had told another, who wanted to sell a twenty-acre parcel in the heart of the roadless area, of my interest in buying. So I took the bait and called the seller. Within a few minutes we'd agreed on a price which seemed to be the going rate and a month later the land was mine. Buying land in the middle of nowhere entails somewhat perfunctory prepurchase investigations. Basically, about all the land has to have is a clear title. Having a boundary survey done would be terribly expensive. Anyway, the East 1/2 of the NW 1/4 of the NW 1/4 of Section 28 is perforce there, and that's what you're buying.

I was aware from the plat book that the northwest corner of the property was along a trail exactly three-quarters of a mile distant from the closest road, and the road at this point was at a 1/4 section mark. Curiosity got the better of me. I wanted to know what I was buying and where it was. Although Jan said, "What's the difference, it's trees and swampland," she did reluctantly agree to come with me and hold the end of the tape as we lined off thirty-nine 100-foot lengths plus 60 feet to reach the northwest corner from the road. It wasn't difficult as the ground was frozen and covered by several inches of snow–just repetitive at 100 feet per line. Along the way we passed a woodcutter sawing cedars for fence posts–a hard way to earn a buck. His astonishment was evident at these people dragging a tape. Part way in we kinked the metal tape, and it broke so the rest of the way was measured at 50-foot intervals. Our method of measuring was not terribly precise, but we did come within 30 feet of an evident mark on a tree, which was the corner. Not satisfied, we then with the aid of compass bearings dove into the trackless interior, a thicket of clasping cedars. I estimated we had fought our way in about 500 feet when Jan went on strike. To avoid a crisis of good nature, it was best to quit and come back another day... with Tom, maybe.

The property had been in the seller's family for two generations. The seller said it hadn't been subjected to any cutting in forty or fifty years. In my treks through the interior it was gratifying never to see a stump.

On an Easter weekend at the end of a relatively snow-free winter, Tom and I decided to cross the entire wilderness area, west to east, off-trail. Nature didn't cooperate. Two days before our scheduled trek it snowed the only significant amount all winter, piling up ten inches. By Sunday the snow was melting, as it only does that late in the season. The only problem was it couldn't soak into the frozen ground, and couldn't run off fast enough. So the cedar swamps became a continuous sheet of surface water through which we sloshed calf-deep, as we slowly twisted and bashed our way along with the compass. Our goal was to come out at the Schloss, a mile and a half later. We thought it would get easier when we reached higher ground, but it didn't. Our route took us through the area devastated by the tornado. Trees lying horizontal every which way create a tortuous obstacle course for tanks or troops, and nonetheless for these two middle-aged boy scouts. (Did I really say that?) We came out of the "bush" where we intended, exhausted and soaked through with sweat, but glad to have done it.

The same neighboring property owner and snowmobile devotee had mentioned that The Nature Conservancy had approached him several times, exploring his interest in creating a conservation easement on his land. "But I don't trust those people," he spouted. It wasn't so much a matter of trust as it was his recognition that a conservation easement would place some limitations of use on his land, which he was unwilling to abide by, being an advocate of "managed forests." "I don't want anyone telling me what I can do on my land."

This managed forest concept is a legislated program administered by the state's Department of Natural Resources and is much touted by the DNR's local forestry agents. Its bottom line is to maximize the long-term commercial exploitation of the state's timber resources for wood products. The governing law rewards landowners who buy into the philosophical premise and are accepted into the program, with a real estate tax break. Though paid lip service to, any benefit to wildlife is purely coincidental. And, it is certainly antithetical to any attempt to protect and maintain significant wildland ecosystems in anywhere near an integral state of being.

A lot of folks think the local DNR forester's advice is next to God's. There are other points of view, which only recently seem to be finding their way into the state's advocacy of forest management. There is an incipient recognition from officialdom that there are legitimate alternative concepts of stewardship that ought to be encouraged, which do not include logging.

To ascertain The Nature Conservancy's interest in these wildlands I called their office in Madison, luckily reaching the executive director of the chapter. It turned out to be an expensive phone call getting acquainted. Shivering Sands Wetlands (their moniker for the entire 3100 acres) was well-known to them, but didn't seem to me to be a high-priority item, likely a problem of not enough money chasing too much need for preservation. Peter gave me a great deal of insight into the problems of preserving this area and of some of the players, confirming intuitions already forming. He also sent me a copy of a preliminary biological survey the Conservancy had undertaken several years prior, to evaluate if this land fit into its scheme of things.

I explained that my goal was to purchase as much land as possible to protect it from any use inimical to preservation. But, my long-term objective was to give it all to an organization with similar goals, a name respected in the conservation world and the wherewithal to provide the permanent stewardship of this gem. I could think of none more qualified to meet these criteria than The Nature Conservancy. We concluded an informal alliance, which I would like to think had potential benefit to each. The Conservancy would much prefer to acquire property by donation than having to make direct purchases, and in turn, I could use the Conservancy's name to help give credibility to my acquisition efforts.

The first twenty-acre purchase wasn't much in terms of size, but it did represent a significant step forward. It gave me some credibility in my quest that simply wasn't possible without owning any property. With the twenty acres in my pocket, I began a letter-writing campaign to the various property owners, inquiring of their interest in selling and letting them know why I wanted to buy. Letters like this usually end up in the wastebasket, so much like pitches through the mail from brokers heralding 10 percent interest rates in a 6 percent market. There's a catch somewhere. No one, but no one, responded in any manner to the letters, not even those property owners who were interested in what I had to say. So I called each property owner whose number I could find in the book or through information. With the owners living locally I sometimes just knocked on their doors. I didn't want to give them the chance to say no over the phone. These follow-ups proved to be essential to getting anywhere with anyone.

Visiting with them on a one-to-one basis helped get them over their initial skepticism and outright distrust, so evident when they realized with whom they were talking. One old farmer I talked with told me he didn't trust "outsiders" and gave me a hard time for a half-hour. This included criticizing my signature on the letter he'd received, telling me it looked like the way "big shots" sign their names. I was patient through the tirade, deflecting as best I could his barbs until he did an about face, talked my arm off for an hour and showed me family pictures. When I left, he said he'd talk it over with his family (wife and eight grown children).

I didn't hold out much hope for the farmer's land. As in any group everyone has a different opinion, which turned out to be the case. One interesting piece of information I did learn was that a sixty-acre parcel he owned in joint tenancy was not with his wife, but with a widowed sister-in-law. It was loud and clear there was no love lost between the farmer and his former sister-in-law. They hadn't even been on speaking terms for a number of years. Her husband had willed the property to his wife and brother jointly, a sure formula for resentment.

At first this seemed to be a more hopeless situation than the family group. But somewhere in the recesses of my mind I had a recollection of an ability to break apart jointly owned property. A call to the attorney confirmed this, so possibly it represented an opportunity. But I had an awful time reaching the sister-in-law, so I finally resorted to leaving a note on her trailer door. Within four hours I had a call. She had received my letter and had forwarded it to her ex-brother-in-law, suggesting that they take me up on the offer to purchase the sixty acres. That suggestion obviously went nowhere. Hell, this old guy liked giving her a hard time!

She wanted out, but had heard that taking it through the courts was very expensive, so never did anything about it. I let her know it not only was not expensive, but it wasn't a lengthy process. I suggested she see my lawyer who was familiar with the deal. She took me up on this suggestion and said she would keep me abreast of things. I didn't need that because I soon got a call from a brother of the farmer letting me know he was going to buy her half of the sixty. I wondered if I'd played right into a trap of my own creation. But the sister-in-law let me know she had only heard about this development through the attorney and wasn't about to let it happen, if she had anything to say about it, which she did. I think it was all a lot of smoke to get me out of the picture. No one in that family was going to come up with the cold, hard cash required to give this substance.

Our transaction took place within days of the court order subdividing the joint ownership. Shortly after the closing, I received a nice note from the sister-in-law thanking me for helping her out of a bad situation; a predicament made all the more frustrating because she needed the money, not the swampland. The thirty acres which became mine included the southwest half of a beautiful wilderness lake.

I found most people to be sympathetic to the conservation ethic, but unwilling or unable to give things away. In almost all my talks with property owners, I heard over and over, "We don't want to see it developed, but..."

"So if you trust the integrity of my purpose, sell the land to me and I'll give it to the Conservancy. This way you will have participated in the preservation of these wildlands, and get paid for doing it." In the current corporate vernacular, a "win-win" deal.

Two adjacent forties were owned individually by two brothers with a common address in Nevada. The first name on the phone number was neither of theirs. It turned out the address for tax purposes was the boys' parents. With a great deal of persistence, I finally connected on the phone. Their father and I had a nice conversation, and he thought both his sons would be willing to sell. The land had been willed to them by their grandfather. Dad asked me to be patient as one boy was in Texas, and the other's home was in Germany. The latter was a U.S. Navy officer on exchange duty on a German destroyer, not due back to Wilhelmshaven for two months. We concluded the first sale before the second brother even knew what was going on. With the ship at sea and then the delays of overseas mail, the second transaction took four months.

One of my unsuccessful attempts involved a Polish lady from Chicago. Having sent the initial letter, I hunted her down at home because she did not have a listed phone number. Getting out of my car in the ethnic neighborhood I saw two women leaving the two-flat. I called to them, which resulted in a questioning sidelong glance and an increase in speed, walking away from me. I suppose there is so much latent fear in these old inner-city neighborhoods that distrust and suspicion are perceived to be paramount to longevity, resulting in the residents' defensive behavior. On the other hand, I don't think I looked particularly sleazy or disreputable in my business suit. For some reason I thought the older woman was my seller and introduced myself to her when I caught up with them. She could not speak English. The younger of the two said she was the owner of the property and was not interested in anything I had to say. She followed this quickly with, "I'm keeping the property so the animals have somewhere to go when there is no other place."

Well here's an opening, I thought, explaining that the Conservancy was a large nationwide organization committed to preservation of wild lands for reasons basically identical to hers. This was lost on her, because she'd never heard of it. Rather rapidly, as she was extremely fidgety, I suggested that we as individual landowners could not succeed in any meaningful preservation, because everyone has differing ideas and goals. For example, if the quality of the land was destroyed on all but her forty, that forty would have little remaining value for preservation. The land needed a common umbrella of protection and stewardship to prevent a piecemeal degradation of the inherent natural systems. From the look on her face, I might as well have been describing a nuclear reaction. Meanwhile, Mama continued to impassively stare at me. When selling something, whether it be architecture, brushes or conservation concepts, there's a time to leave, and it was overdue. While saying a polite goodbye to the two women, I saw unchanged the same distrust evident after my introduction. As I walked back to the car, I thought to myself that I hadn't even made a dent. I did send her three copies of The Nature Conservancy's magazine, which clearly states the organization's purpose, so at least it might gain a degree of respectability in her eyes. In an accompanying note I suggested that if she didn't want to sell to me, she should sell directly to the Conservancy.

This has not been my only rejection by any stretch. The circumstances were simply unusual and the attitude obscure. I've always wondered what the real story was.

Other's attitudes were becoming clearer. Coincidentally, my friend and fellow landowner (on the snowmobile) called me in the fall and excoriated me for having bought the property containing the lake. "Why I've been trying to get that land for years," he ranted, "and now a damned carpetbagger grabs it." Without taking a breath, he then asked if he could keep his duck blind on my land, to which I acceded. He also asked if he could improve one of the two trails leading to the lake through my land. I'd been on the trail, and it was not easy walking. But this man doesn't like walking. His means of travel, when the snow's not the surface, is a four-wheeler. Improving this trail would have meant he could have virtually stepped off his ATV into his blind, rather than schlepping gun and decoys, cushion and coffee for a couple hundred yards through the cedar swamp to reach the blind from the other "improved" trail.

I luckily replied, "If there are going to be any trail improvements on my land, I'll do them myself. Then I won't be disappointed in the results." At this point in time I wasn't even aware he owned a small bulldozer.

I was aware there was one of these around, because the prior year someone had driven one along a rough trail running a half a mile through my land to reach a tree blown down and uprooted. The tree blocked passage along the path, not for walking or skiing, but for any motorized means of travel, such as a snowmobile. This impediment didn't bother me at all, but it obviously did someone, who had the audacity to force his destructive way to the tree to shove it and its root wheel aside, and then turn the equipment around through the trees to bulldoze his way back out.

Fortunately, there was deep snow cover over frozen ground so the damage to the humpy, tussocky underlying surface was minimal, but the cedar trees flanking the trail took a beating. It wasn't the damage that bothered me as much as the principle. Though the downed tree was not on my property at that time, the access route to it was. No one needs to ask permission to walk through, but I felt that grinding over the path with a bulldozer is quite another proposition, indeed!

The downed tree was located on ninety-six acres adjacent to the extreme east boundary of my land. The west half of the ninety-six acres is a continuation of the dense cedar swamps. Walking eastward toward Lake Michigan, the terrain rises up to a beautiful upland forest of hemlock and white pine with a sprinkling of beech and birch. It's uncanny that upon reaching the more open upland forest, the soft drone of distant surf becomes audible over one-half mile inland from the lakeshore. This land is contiguous with the cutover 110 acres on the south, and had a four-acre chunk, carved out of its northeast corner. This tiny parcel amidst the forties and one hundreds puzzled me.

There were three trails on the higher, drier elevations; breathtakingly scenic tunnels through the hemlocks. Though a certain amount of cutting had been done in the groves of these trees, it had been very selective. Except for the lopped off tops laying around forever, the visual impact of the logging had been mercifully low.

The Conservancy's biological study had indicated the presence of a large bog of ecological significance somewhere in this area of Shivering Sands Wetlands. In reading the biological description and studying the directions given in the report, I was almost certain that the bog lay concealed on this ninety-six acres. A January day with only a half a foot of snow cover, presented an ideal opportunity to search for it. Compass in hand, my search on foot proved out. There could be no mistaking the singular natural feature that I was standing in the middle of for that described in the study; a well hidden, yet expansive clearing in the surrounding forest. From the air it must be obvious.

The diversity of the topography and the natural features of this property are enhanced by its strategic location in the entire wetland. Its almost undisturbed forest was an undisclosed gem. The stewardship of the current owner seemed one of benign neglect. Ominously the east (and higher) half of the property was zoned residential. The boundaries of the ninety-six acres reached deep into the interior of the wetlands; a keystone for comprehensive preservation, or conversely a salient of development cleaving apart and subdividing the large flanking wildlands. Importantly, it is adjacent to some of my land. Access to one is from the other on a single path—that of the downed tree.

I had a vested interest in this land, so I devoted some time to research its history of deeds in the county recorder's office. The lack of accessibility to it from a road was the singular feature of this property that had kept it from development. To reach it from the Drive meant traversing the length of a dedicated right-of-way 400 feet long and only 25 feet wide, still owned by the original land developer of the residential lots along the Drive. As if this wasn't problem enough, one had to cross the entire width of another fifty-acre parcel and several hundred feet of the logged off 110 acres to reach it. Access across the fifty acres was guaranteed in its deed, but there was no defined easement. A wheel track in the earth indicated the route was used primarily by hunters, loggers and poachers. The local sawmill operator owned the intervening fifty acres. He's a brother of the owner of the 110 that had been for sale for years. He wasn't the logger that ruined the 110, but he was doing a slow but steady number on the fifty to keep him occupied at his mill down the Drive.

As I've become acquainted I've found, surprisingly, that everyone seems to be related to everyone else... somehow; a facet of small-town life foreign to anyone not raised in this culture. My awareness of this phenomenon began at PJ's and seems to be ever unfolding. Roger, my errant plumber, cum electrician, is somehow related to Jim, and Jim is to Debbie, and Debbie's husband is Greg's cousin, and on, and on. The web of familial relationships and their genesis never ceases to amaze me. I can't recall how many times I've heard, "Oh, he's my cousin," or "My mother's sister is married to her father," or some other such incidental introduction to these interlocking shirttail ties.

The president of PJ (and coincidentally my roommate in college) had been elected to the board of directors of The Nature Conservancy. We talked at length about Shivering Sands Wetlands and specifically about the importance of this ninety-six acres to the wetland's preservation. We decided to visit with its owner, a somewhat crippled and retired local contractor. Bill was representing the Conservancy, and I myself, to gain some insight into the owner's intentions, and to determine if he would consider selling. He said he and his son were investigating the possibility of developing it. This was really curious as fifteen or twenty years prior he had concocted a half-baked development scheme, but had only managed to find a pigeon to buy one parcel; the four acres out of the northeast corner. During our meeting in his home, it was obvious his wife and son, both listening in on our conversation, were adamantly opposed to selling, and were even antagonistic with our visit and with our offer. After leaving, we conjectured that the unfriendly wife was afraid her husband was going to give away the store.

I felt the passage of time since the original attempt at development had only made immeasurably more difficult and expensive any current development scheme. Environmental restrictions and regulatory requirements governing developments and land use had in the interim grown from meager and unenforced, to sophisticated and costly obstacles, which shock and discourage uninitiated would-be developers.

"Even if he is serious about this, his development concept is a pipe dream which is just going to take some time to burn out," I commented to Bill after the door shut behind us. It was going to be one of those instances where I was going to have to "cool it" for awhile, because his sale price, in lieu of developing, was as unrealistic as the subdivision idea. The property had intrinsic value to our goals so it was tempting, but very unwise to capitulate to the high price. Such rashness would set an undeniable value precedent for all future purchases of similar property.

We did learn they had discussed developing the land with the county planning department, once! We wanted to keep our fingers on the pulse, but it was tough following up on our initial visit. My strategy was to call when I thought the wife might not be at home, as a letter I sent had suspiciously never reached the man. When the wife answered the phone and learned who it was, I could feel instant tension at the other end of the line. I was always curtly informed that her husband was very busy and couldn't be disturbed. He never returned a call, and I'll bet he never knew about the calls to do so.

Involved with other things over the months, my active interest subsided to a latent hope that my perception of things wasn't wrong. I'd heard he was going to sell to his son. So, it was with great surprise that I got a phone call at work from the erstwhile seller well over a year after our visit. He asked if I was still interested in purchasing the land.

I said I was if the deal resembled the offer previously made, and reiterated the terms. He replied this was OK with him. It took about a microsecond to realize that I'd better move fast. "I'm going to be in Sturgeon Bay on Saturday, and I'll have a real estate sales contract and a check for earnest money." On Saturday afternoon they signed the contract and took the deposit. "What changed your mind?" I asked.

"Our granddaughter wants to go to college and we're the only ones who can help. Her dad died several years ago." A truly generous and commendable gift to the girl from obviously loving grandparents. I said my local attorney would be contacting them and left it in his hands to close, fast! But it didn't turn out to be so easy with "no shows" at appointments and unanswered phone calls. After the transaction took place, the attorney commented that he wasn't at all confident it would take place right up to the minute they signed on the dotted line and accepted the check for the balance due. The wife apparently was fighting the deal to the bitter end. I have the impression that she was never satisfied with the price. The phone call from the attorney telling me it had closed relieved a lot of anxiety. I think I must have asked him three times if he had recorded the deed.

This was a significant step forward, doubling the acreage I owned in a single purchase. I felt such exhilaration when walking the paths through these forest glades. The huge stumps of the original forest remain just discernable, protruding up from the forest floor as rotting monuments to the original commercial exploitation, so total that not one tree remained. These stumps are often hosts to newer trees growing right out of the decaying mounds. But no longer would these second generation trees cringe at the sound of a chain saw, nor fall to their bite.

Everything stayed the same until early the following winter. I'd asked Greg to go in with small equipment and remove all the evidence of logging; the leftover cut logs, the treetops and slash along the trails, the only places where any of the now large second growth had been cut. On their way out they erected a gate of sorts at the entry to the property along the track emanating eventually out onto the Drive. I designed a gate of wood posts cut from power poles purchased from the utility company. These were spaced to keep snowmobiles and ATVs from using the trail. Two in the center were turned down by Jim on his lathe to fit inside ten-inch well pipe sleeves bedded in concrete.

These posts were fixed in place with padlocked bolts through the piping. Unlocked and lifted out, these removable posts provided passage for equipment should we have need to get in again. I knew it was an honest man's gate, but felt it and the sign would give pause to all but the most determined. I just didn't know how determined determination could be.

The Conservancy and I signed a cooperative maintenance agreement covering all the land, and they had several small signs designed and painted for placement at each of the trail accesses. These signs stated that the property is managed in cooperative arrangement with the Conservancy, and published the rules of use, most significantly that no motorized vehicles were permitted. Within days after Greg posted the first sign at the gated entry, I had a call from my now neighboring landowner and snowmobiler obviously upset with the gate and the prohibition of motorized vehicles.

I explained that the Conservancy does not permit motorized vehicles on their properties. Although this was not per se Conservancy property, the management agreement covering the land certainly was a reflection of their policies. "If I didn't subscribe to these precepts, there wouldn't have been any agreement to start with. You may find this an inconvenience, but I can't start making exceptions before the ink dries. Should you be an exception, your snowmobile track past the prohibitory signage would be stark evidence of disregard for the posted wishes of the property owner. Why would I go to the trouble of an agreement only to subvert its effectiveness."

He must not have realized the seriousness of my intent and thought he'd test the waters a little deeper. "Well, how would you feel if you ran across me driving through your property?" as much as to say, is it OK if I don't get caught?

The only reply to this had to be an unequivocal and direct answer. "Most particularly because of this conversation, if I should meet you on your snowmobile or four-wheeler on the trail, you will know an unhappy neighbor. If nothing else, use of these trails by mechanized equipment spoils my enjoyment of my land. I don't want my ski trail prepared by a snowmobile, nor torn up by the tires of an ATV churning up the snow down to bare ground."

This was the end of any reasonable conversation. "Why these trails are for everyone to use. The property owners have had an 'agreement' to this effect, and I've maintained all these trails for years." (BINGO!) And becoming ever more indignant, "I have a right to use them, and no one is going to keep me from using them as I've done for years!"

I don't like confrontations, but I like ultimatums less. "Wait a minute, you know very well that there has never been any agreement between property owners. Some have never even visited their land, others live in other parts of the country, and the rest don't give a damn. You're the only person who uses these trails consistently, and any 'maintenance' you've done has been for your convenience to use the trails as you see fit."

My affable neighbor was sputtering out of control and hung up. Whew! I had no desire to have an adversarial relationship with this man, but it was certainly evident that giving in would be giving up the use of my land in harmony with my goals. My own enjoyment would be subordinated to someone else's inconsistent vision of how my property should be used.

This turn of events was upsetting and left me uneasy, suspecting that this wasn't the last I was going to see or hear about this basic philosophical conflict. I just had no idea how embittered my protagonist was to become, and how much mutual frustration was to be generated over time.

During the first winter of my ownership, someone on a weekly basis was driving their snowmobile over one of the trails with total disregard of the signage. Two of the timber gate posts were just far enough apart to squeak between, scraping the sides of the machine against the posts. It wouldn't have mattered if the posts had been closer together as whoever it was would just find their way around. Skiing on Saturdays was exhilarating, often through unbroken snow. But Sunday mornings, the snowmobiler or ATV jockey got there first.

The trail used most often by the machine is the one with the post gate near the southeast corner. Its course bends around and leaves the property near the northeast corner. I really needed to acquire the four-acre chunk, and tried, but to no avail. My letter to the address in Ohio, which was recorded on the real estate tax documents, was returned with "No forwarding address" stamped on the envelope. I spoke to the town treasurer (the tax collector) about it when paying my taxes. He had had the same response on the real estate tax bill. The following year he had a new address to which I communicated. The owner was an attorney who twenty years prior had enjoyed vacationing in Door County. He loved the area so much that he became the sole pigeon in the half-baked development scheme which went nowhere. He said his family had grown up and had their own vacation idylls, and he and his wife had bought a place closer to their permanent home. He was so effusive about the "beautiful four acres," I hadn't the heart to tell him that his land was almost all swamp. Apparently, what he had thought he had bought was not what was actually his. From other evidence of old survey marks, I have always wondered if the former owner, and erstwhile developer, had erroneously plated the land. It had all the appearances of a do-it-yourself operation. At any rate, within minutes we reached an acceptable price, and the four acres soon became mine, filling out the missing piece at the northeast corner of the one hundred acres.

Further evidence of inaccurate surveying soon came to light. I wanted to erect posts and a Conservancy sign on the trail at this end of the property, but had only an inkling where the east line lay from using a compass, starting at what I thought was the southeast corner at the gate. It was good I spent the money to have a land surveyor establish in the field the east property line. The results of this survey were surprises. The surveyor told me over the phone that the southeast corner was 246 feet east of where it was thought to be (at the posts), and the gate was several feet south of the south property line. I thought I'd just let the latter revelation remain my secret until someone else discovered it, which was highly unlikely. The original surveys of this area had skewed the north/south section lines 12 degrees from true north.

The previous fall I had explored with a compass what I thought was the east boundary hoping to find a corner post at the northeast extremity. Talk about looking for a needle in a haystack. I didn't find the corner, but did run across an unusual rock escarpment on one side of the swamp. This limestone cliff overhung its base by 10 or 12 feet. The overhang, not quite high enough to stand under, provided wonderful shelter for animals, and evidence of its use for this purpose was abundant. The floor under this rock canopy was carpeted with deer droppings several inches deep–a dry, soft bed sheltered from snow and rain where the deer evidently yard up in the winter.

The posts, set in at this new gate, were placed close enough to preclude any passage through. But, within a few days a determined person on a snowmobile bashed his way around one end through some hemlock saplings. Addressing this we erected several more posts to where the ground became steep and impassable.

Skiing in a week or so later, the usual single snowmobile track was imprinted in the snow. I wondered what I was going to find this time. The evidence at the gate was so clear it was almost as though I had been there watching. The snowmobile track stopped outside the gate posts. Footprints described a contemplative path, searching for a solution beyond the ends of the gate. Then the machine tracked up a steep embankment through some small spruce and hemlock for 20 or 30 feet returning down to the path to continue along, its driver quite smug, having again outwitted the impediments placed in his way.

Another time a friend and I were delighted to awaken to six inches of new fallen snow. With great anticipation of untrammeled snow conditions, we ate a quick breakfast and departed out into the crisp air of a calm winter morning. Breaking the track through the fluff sparkling in the sunlight, we had apparently beaten the snowmobiler. Maybe he was sick. No such luck! He had taken another trail through from the west end where there are also posts. At this entry point the forest is so dense and the width of the path so narrow that going around this gate would be impossible without cutting. We decided to ski out to the gate, a mile away. It was discouraging along the trail. I am trying to stimulate and promote the growth of new trees coming up along the trails, but they don't have a chance with the snowmobile pushing the fragile stems over, often breaking them, and tearing off the foliage under its tread. This is paralleled in the summer with the use of ATVs. I don't often get into the forest during this season, but two years ago there were very few bugs, and I took advantage of this unusual phenomenon, walking along the paths, enjoying the diverse character of the summer forest. I discovered some unusual orchids growing low on the path and took care to not step on them on my return. Two weeks later I found them mashed in the tire marks of a four-wheeler which had passed twice along the trail, it being a dead end.

The mystery of the snowmobile entry was solved as we approached the west end. The center of the three posts was gone, as was the Conservancy sign. How long they had been gone is anyone's guess, except to the person who removed them. My companion said this insistent intrusive violation of another person's property reminded him of stories of the Old West. It's like the range wars between the cattlemen and Johnny-come-lately farmers, whose fences were destroyed as fast as they went up by the ranchers who wanted to keep "their" range open, not willing to recognize nor accept legitimate rights of ownership, except theirs, of course. We skied that morning along the singular snowmobile track with purpose, deciding to follow it to its source. The reader can only guess to whose yard it led, admittedly proof positive of only that day's incursion. But the word always gets around. Upon being introduced to a man at a local conservation meeting, he shook my hand and said, "Oh I've heard a lot about you from your neighboring property owner," and broke into a grin like that of the Cheshire cat that ate the canary.

Early this year, the farm couple with the 110 acres was getting more serious about selling. Previously, I had asked the Conservancy's property acquisition person to visit these folks to demonstrate its interest in their land, and its seriousness of intent with respect to the preservation of the entire wetlands. The Conservancy's representative felt she had developed a good rapport with the farmer and his wife. The farmer "toured" Kim through the property, going so far as to comment that he would like to see the Conservancy have the land, but his wife was not so inclined. Unfortunately, it was abundantly clear who called the shots.

I received a letter from the wife stating that they had reduced their price, and the first person who met it would get the 110 acres. Intuitively, I felt that this time she meant business, and I called the Conservancy. They had also received the letter and were contemplating what to do about it. At this point in time it was inopportune for me to come up with the considerable amount of money to undertake the purchase, which was unfortunate as this land was a linchpin in the entire wetlands preservation strategy. Its loss could have serious consequences as the land was zoned residential, and was probably developable by the right organization or person. It also lay astride the trail access from the Drive. The Conservancy's leaders equivocated, and I came down directly, pointing out that I had undertaken this venture in land acquisition because of its stated interest in preserving these wildlands. I felt this was the time for the Conservancy to demonstrate its dedication to this project, or much of what I had accomplished would have been for naught. They moved fast. Kim arrived at the farm the next morning with a formal offer to purchase priced at $1000 above the couples asking price, a clever and fortuitous ploy. By 10 a.m. the deal was concluded. That afternoon two more offers came in, one at the asking price from a party apparently interested in developing the land.

It was time to let the Conservancy know it was buying my gate and that if they wanted it moved, I would do so. As our management philosophy was identical, it didn't make any difference to them on whose land the posts lie.

The Conservancy was aware that the farmer's brother (the sawmill owner) with the fifty acres between the Drive and our lands wanted to sell when the farmer did. But it didn't move fast enough. The fifty was sold in a flash to some local people. In the Conservancy's defense no one could have foreseen the degree of imperative—to move on the fifty almost in lockstep with the 110.

It appears the new owner is sensitive to the concerns of preservation and, in fact, has erected a formidable gate to prevent motor vehicle use. This was becoming critical as there has been ever increasing use by people finding their way back into the Conservancy's property with four-wheel-drive pickup trucks, jeeps and the like. What had been simply a tire track in the grass was becoming deeply rutted in places. Though the owner thoughtfully gave both the Conservancy and myself keys, I'd rather walk. The only time any motor vehicles have entered since the gate was installed was when I asked Greg to replace a post that had been lifted out during the summer, and to wreck and haul away an old tarpaper hunting shack on the Conservancy's property... along with all the junk laying on the ground around the shack. We like to do this sort of maintenance work during the winter when the ground is frozen so there are no impressions of tire marks left behind as reminders.

In the middle of the holding tank crisis I received a disturbing letter from an attorney representing the town government. It stated that the town board had complaints from "neighboring property owners" that I was obstructing town roads in sections 21 and 28. The letter demanded that I remove the obstructions forthwith to permit the public to have access along these roads. In so doing I would avoid legal action being taken against me by the town. A copy of the letter was sent to the town chairman. I couldn't fathom why the town would be getting involved, as none of my land in these sections came anywhere near any town right-of-way. But it was aggravating as I felt I'd better hire an attorney to find out what on earth was going on and to ensure that I was on solid ground.

So the attorney would have a better awareness of what these "roads" were, I asked if he would put on a pair of old pants and high boots and walk with me along these trails. Those two hours we spent trudging through calf-high water on the swampy paths on a Sunday afternoon in late spring was a hard way for him to earn a fee. He finally gave up trying to keep dry and plowed along with mud-filled boots and pants soaked black up to the knees. Each step was a struggle with the mud trying its darndest to suck off a boot as it was slowly lifted out of the glop.

We had to address two separate issues. The first was the claim that these were town roads, which was preposterous and totally undefensible. Simply enough, the town had no right-of-way. The second claim was the right of other landowners to use these trails on my land as they see fit, but this had nothing to do with the town government. There are instances where people have claimed prescriptive use which entitles them to continue their longstanding activities on another's land. But this concept is well-defined in law, and this circumstance met none of the criteria.

I had faxed the letter to The Nature Conservancy, which now had a direct interest because of its recent purchase of the 110 acres. Any attempt to force me into defacto giving away my rights as property owner would place the Conservancy in the same position, as the trail also ran through its land. On a much broader perspective, it could have profound implications for the preservation goals throughout the entire wetland. The Conservancy offered its attorney's aid, if required. Upon reflection, we all agreed it was a terribly clumsy, ill-conceived threat with absolutely nothing to give it any credence, nor force of law. It was curious that the lawyer who wrote the letter wasn't even the town's retained legal advisor. It's also strange that the town chairman would permit himself and, in effect, the town board to become involved in what is nothing more, nor less, than an imbroglio between two individual property owners.

My attorney called me after having spoken to the other lawyer. He said, "You know, when lawyers are aware their position is one of strength they are not reticent to let their adversaries know. Even if their position is not terribly strong, intimidation is part of the game. This was the most one-sided conversation I have ever had with another attorney... with me doing all the talking! I don't think you'll ever hear about this again." But, the intrusive use continues. The intimidation tactic simply harkens back to the ranchers utilizing their membership in the good ole boys club to threaten the farmers. Kim says that sometimes effective preservation has to await the passing of a generation, hopefully to a more enlightened one, following. That's a scary premise at the rate wildlands are being eaten up.

## Chapter 24
# Autumn

Fall caught up with us on *Aurora's* annual cruise with two storms of gale-force winds in the last five days. The first came out of the southwest and the second smack out of the north with nighttime temperatures dropping into the thirties. At least the north wind provided a sleigh ride southward to Sturgeon Bay as it moderated. It seemed an age since I had been home, and now fall got there before I did.

I had been accumulating dirty laundry for the conveniences of the Schloss. While the laundry washed I enjoyed the luxury of a good shower and then walked slowly through the woods saying hello to all my friends. And I compulsively picked a few of the many weeds emergent in the junipers during my month's absence. The house seemed empty without Muggs. How symbiotic both are to my spirit.

Sailing a boat and living at the Schloss each pull me in their own antipodal direction. Each is ultimately rewarding and challenging in its own way. Each blesses my soul with relaxation and communion with the natural world, and each requires attention. Both must operate well and be maintained in mint condition. Things left undone or unrepaired nag at my enjoyment. In a way it's too bad that I can't be like my friend Tom with the beautiful cabin in the U.P. The place is falling in over his ears in desperate want of care. He, with carefree blinders, loves it, and uses it and abuses it year after year. I'm the nag– "When are you going to rechink? It's getting like a sieve."

"I've got to get at that next year."

After thirty-five years of friendship, you'd think I wouldn't bother because I know what his reply is going to be before I ask the question.

My feeling about *Aurora* is that first we have to make it look like a yacht and then we have to learn to sail it well. Only then can we ultimately appreciate it for all its worth. At the Schloss a similar feeling motivates my activities. Mentally, I began to make my notes of fall projects, some deferred from summer's aversion to toil. Within two weeks the enduring contest for my attention would be suspended with *Aurora* decommissioned and dormant in my mind for a couple of months. So my anticipation of returning to and being at the Schloss with regularity was becoming almost overwhelming.

High on my agenda was to visit the conservancy area, through which I had not walked in three and one-half months. Besides the sheer pleasure of being there, I get concerned about what damage it may have suffered from insensitive use during my absence. Fortunately, much of the roadless land is too wet and the forest too dense for anybody but the most stalwart to penetrate even with a compass. But where there are trails, there is frequently evidence of careless use and, sometimes, downright malicious destruction.

During July a biologist doing some field work for The Nature Conservancy had let his headquarters know (and they in turn me) that someone had torn out of the ground one of the barrier posts preventing motorized vehicle use and had been into one of the trails with a jeep. He had gotten the license number of the parked jeep. As luck would have it, somewhere between the biologist and me the number was repeated wrong as it was investigated and found to be registered to an Oldsmobile in the southwest corner of the state. Other than the displaced post, the property survived the summer in its pristine state. No beer cans, no tire tracks of ATV's bashing through the underbrush circumventing gates, nor trees mutilated or destroyed.

It is difficult to imagine, or at least it was for me, that someone would thoughtlessly, or with malice, damage or destroy living things on another's land. My innocence in this regard was raped away one afternoon several years ago, shortly after I had acquired the land Jan and I were walking a three-mile circuit when we heard a distant chain saw grinding away. Puzzled we stayed alert to the presence or evidence of anyone having been on the trail. About a quarter mile in from the end a 16-inch hemlock lay tilted over the trail, its branches hung up in those of other trees. Its trunk was newly severed, the sawdust plain around its stump. Tire tracks led back down the trail. Since our presence was not known, the vandals must not have cut the tree for the wood but only for some warped pleasure. Maybe they were just trying out a new saw. I left the forest that day saddened with the senseless loss of the beautiful tree and angry over being victimized by this contempt of another's property. It was then and there I made up my mind to prevent access by any motorized vehicles on this trail; the only one that is anything other than the narrowest footpath. But, the reader already knows the rest of that story.

So I walked along the paths this autumn day with a sense of relief which permutated to one of harmony with my surroundings. I soaked up all the sensory and spiritual pleasure of walking through the undisturbed forest on this beautiful day in the month of October.

The path winds through deeply shadowed glades of hemlock and pine, the understory still dappled with sunbursts of maple and beech. Underfoot lay a thick potpourri of autumn's gift to the soil. The matted grays of last year's carpet have been blanketed under this season's splashes of yellows and burnt oranges. The bright hues of the newly fallen leaves quickly fade to siennas. All shriveled up and dried to a crisp, they make each step as noisy as eating a bag of potato chips. An occasional crimson leaf is a vivid complement to the ferns and princess pines and raspberry leaves, which are the only remaining greens on the irregular duff. Resting on a bed of needles in a patch of sunlight just off the trail, it felt like summer—too warm for a jacket.

I lay on my back looking at the blue sky through the tops of the birch and maples swaying in the wind; a wind that in the shade blew cold and made my nose run. But prone on the pine needles only an occasional puff wafted this low to remind me that this wasn't summer. I felt a sense of peace and contentment with Muggs tucked against my side, ever alert to her surroundings. Her nose probing and twitching and ears swiveling—a study in ancient, but unforgotten, behavior. The chatter of unseen birds became intermittent, and the wind aloft in the trees abated with the sun. At times the only thing audible was the occasional scratchy noise of a leaf parachuting lazily down through the nearby foliage. It's surprising how loud this incredibly small sound can be when there is no other.

A sudden tension in the black body next to me broke into my reverie. It's always a signal that something or someone is afoot upwind. As the tension mounted within Muggs, she quivered and whimpered once or twice. I saw the source of her concern before she did. Coming toward us along the path were two does. Stopping, looking and browsing, they were enjoying the sunny day so much like two women strolling through a park, pausing to pick a flower or two and chatting away a pleasant afternoon. The protracted stress of this was getting to Muggs. How long could she possibly stifle a bark? Whether from my whispered words of calm, or from a realization within that we were the intruders in another's territory, she kept silent, watching with rapt attention the approach of the deer.

When just abreast of us, one of the does looked directly our way and realized something was amiss. When the second stopped and looked, Muggs broke the spell with a soft half bark–half growl–what I call a browl. It triggered instant flight. Two white tails bobbed away as the deer dodged and weaved through the trees into the obscurity of the forest. From this must be derived the term "hightailing it." The brave Scottie wanted to give chase which, with eight-inch legs, would last at least twenty feet, when the required display of bravado petered out as the distance between us increased.

If nothing else, this encounter illustrates that to see wildlife up close it helps to settle in and become as one with the surroundings for awhile. Then the doors start to open in what had seemed only an empty forest. Usually we plow our way along the leafy paths, motoring away with our mouths. Our intrusive presence is telegraphed ahead to all creatures who make themselves scarce, or watch silently until the threat has passed. Two exceptions may be the porcupine and the partridge. If unlucky enough to be caught on the trail, the porcupine has its armor of quills which is a formidable substitute for this creature's ponderous waddle. The partridge stays hidden in the trailside brush until I practically trip over it, when unexpectedly it explodes out of its cover to leave me frozen in midstep, needing a moment to regain my composure. Even if you're in the woods hunting for them, their violent burst into flight is nonplusing. It takes a lightning-like reaction from a cool head to bag one before it rockets away behind the trees. If you're deep in thought or reflecting on other things, the bird is out of sight faster than an expletive is out of your mouth.

\* \* \*

The cooler weather brings out the more serious cyclers. In their black spandex tights, hunched over the handlebars with total sense of purpose, they zoom along the curvy Drive like wasps going for the kill. They travel in packs so tight that a mistake in judgment or a sudden stop by the leader would put pedals into spokes, causing an instantaneous jumble of bikes and bodies. Summertime cyclers are there for relaxation and the scenery, often stopping to rest and have their lunch at the overlook. Not those of fall. They never seem to look anywhere but dead ahead and they get their nourishment on the fly from a bike-mounted bottle. Maybe that's why they all look so emaciated. It's curious why these Grand Prix types bother to spend the money and undertake the long drive for a weekend in scenic Door County when they could achieve the same sense of speed and competitiveness on an indoor track. Of course, I have the same wonder about cross-country skiers who sprint their way along a trail at Olympian speed, never stopping to "smell the roses"... so to speak. Conversely, if these competitive zealots saw me picking my weeds they'd probably be thinking I was doddery. Oh well, everyone's doing their own thing in their own way. I'm just glad they don't all like my thing... and I don't mean picking weeds.

It is time to pluck the geraniums out of the flower boxes on the deck. They've gotten frostbite and the deer have nibbled away the leaves on the plants in the pots along the walk. At least I assume it was the deer, since they also did a real number on the hosta lily leaves, and it took a prodigious appetite. Muggs thinks it's great when the winter covers get put back on the flower boxes. She uses them for elevated perches from which she scans her world and naps when her eyes get too heavy from the duty.

One afternoon she caught sight of a skein of Canada geese approaching along the coast. They banked inshore directly toward the house and passed overhead at rooftop height, to disappear from sight on the back side of the ridge. I didn't know which to look at— the geese whose movement was so slow they appeared to be in suspended animation, or Muggs with her neck craned upward looking with unblinking curiosity at these monsters floating in the ether. They were so low I could hear the air slap with each downbeat of their powerful wings. God, I'd love to have known what was going through Muggs' mind as she so intently observed this airborne parade. It always sends a shiver down my spine when I see and listen to these magnificent birds in their disciplined, yet fluid, formation flying.

# The Paradox of Paradise

Their calls often reach our ears at great distance. It takes a few minutes before we can see them high in the blue infinity. I never cease to be awed at the wonder of it all... this grand display of the natural world in working order.

It reminded me of an October canoe trip with Tom years ago in northern Ontario. The noise reached us first. We stopped, and with the paddles across our legs, the canoe drifted while we watched approach what seemed an endless flight of geese in multiple vees. They wheeled through a broad circle over the lake, boldly silhouetted by sheer number against the fiery twilight. In the distance their noisy honking was only a cacophony deafening in the solitude of the calm evening. When descending close overhead, the birds' calls became distinctly separate. They seemed to be voiced with purpose and order, as though requesting in-flight adjustments by their wingmates. These calls were responded to by a slight dip of a wing for control, or a single sweep for momentary acceleration or lift. They landed swish after swish on the orange lake some distance ahead until there were thousands of birds crowding the surface. We had been silent during this prolonged landing, and we remained speechless for some time after in profound reverence of this sublime wilderness spectacle which we had experienced.

\* \* \* \*

Early in the fall the bees are pesky on the deck as they go about their business from flower to flower. Come to think about it, the bees are even pesty when sailing in this season. They disturb the helmsman's concentration and they skim down below to innocently threaten 220-pound deck apes, turning their macho cool into paralyzing fear.

The monarch butterflies, those champions of international flight, still grace the meadow in front of the house. At first glance their flying seems aimless. When observed for a while, it is evident they are quite absorbed in their quest for nourishment. This fall we gave a monarch a ride back to land from thirty miles offshore. It latched on to a seam on the mizzen and never moved until we were within a quarter mile of our destination. Then with a pirouette and a dip, it took its leave from our rig to continue again its long journey. There have been far fewer monarchs around the house than in past years. Probably the result of an unusual frost a year or two ago, which killed them by the millions in their winter nesting trees in Mexico.

\* \* \* \*

The errant painter turned up the last Saturday in October to finish the job begun in June. He thought because it was a rare 75 degrees in town he'd have a great day to wrap up the last coat on the large bay. It was 51 degrees on the coast with the surf pounding ashore driven by the same south wind that over land blessed Sturgeon Bay with the unusual summer day in late October.

Keeping paint on this bay's trim has become a perennial refinishing problem due to the extremes of climate. Nothing seems to work for long. Could it have been a design fault of the architect? After having consulted with everyone who ought to know about these things, I am no closer to an answer than I was before I started scratching my perplexed head. The architect did everything he ought to have done– I keep telling myself.

Shivering in his short sleeve shirt, the painter, shaking his head, said, "It sure isn't a low-maintenance house." If low maintenance equates with aluminum siding, I guess he's right. On the other hand, the cedar shingles just do their thing year after year turning from cinnamon to ultimate gray. The shingles on the side of the house facing the open lake have turned totally gray, the other sides have undergone lesser degrees of fading, depending on their exposure to the sun. Within the log alcove, which is covered by firewood much of the time, it's practically the same as when built. Admittedly, there is an inordinate amount of white-painted trim. The lakeside trim becomes streaked with dirt only a couple of months after it's washed. The onshore winds and extremes of temperature, both hot and cold, and shoreside damp and sun dry, blast this side of the house unmercifully year-round. But, I wouldn't do it any different if I had to do it again.

If he gets the job finished, it will be miraculous as suitable days to paint in this environment this late in the year can be counted not on one hand, but on one finger... maybe. For some reason or another, every day that is good he always seems to be in the middle of a major wallpapering job; or, its always rained on the days he was going to do it. A real flexible scheduler, this man. One wonders if the color of my money is different. Good God, if this customer was treated well by a painter, the painter could secure a lifetime retainer working on this high-maintenance building.

Painters come and go. My luck with this trade has been poor at best. The painting subcontractor used by Yukon was probably the only disappointment of the building experience. Ted from PJ's, who did the miscellaneous painting, shingle dipping and table finishing, went out on his own with reasonable success, but decided after a couple of years of being his own boss, he'd rather manage a mushroom farm. I hated losing him to mushrooms as he was skilled, meticulous and above all reliable.

The next painter did a good job, but his assistant used the wrong white paint on part of the exterior trim. It had a pinkish cast. He wouldn't believe me, so he brought out a sample of the paint I had given him to use—to his embarrassment. When proven wrong, he (or his assistant) repainted it, and he charged me to do it over. In our business when we make a mistake, we rectify it at our expense. He was incensed that I wouldn't pay the bill for the re-do. When I called the next year, he told me in a cursory manner that he had retired. It turned out that it was simply his way of telling me to get lost. I think I'm missing something. I should have been the angry one. Besides the aggravation of having to search for another painter, I wasn't disappointed at the breakdown in the "relationship." Knowing how long things take to do from past experience, I always had the feeling that I was being overcharged.

In our conversations there always seemed to be an attitudinal undercurrent of resentment toward those people who could afford to employ a painter rather than doing it themselves. Hard to fathom, but possibly indicative that retirement really was overdue.

But that's history, as they say, because I'm one painter beyond and looking again. Recommendations are hard to come by, as everyone I know around here does their own work. It's too bad there aren't search firms (those clearing houses for executive talent) for painters and decorators. In today's marketplace, executives could only wish that their expertise was in as great demand as painters seem to be. No painter from the Sturgeon Bay area even bothers to be listed in the Door County Yellow Pages.

\* \* \* \*

The last day of October was the first day it really felt cold. Fall lay on the ground. Only the remarkable maple tree outside the kitchen window still had its leaves. The first snow blew in on express squalls, lashing short duration blizzards of white pellets through the barren trees to land on the dried leaves with a crinkly hiss.

It dusted the Bar Harbor junipers; their summer green faded to mauve, a sure sign of winter just over the horizon. I was splitting cedar logs for next year's kindling and didn't notice the raw wind once I got into the rhythm of the job, except when I paused for a breather. Only then did my hands complain. It feels clumsy when I use an axe and a sledgehammer wearing gloves. Surehandedness, a good eye and focused attention are essential to one's well-being when swinging these lethal weapons.

The distant ribbon of coastline appeared purple, separating the mottled leaden sky from the teal water. Out near the horizon, the tops of some seas turned sparkly and brilliant, electrified by the sun streaking through a momentary break in the clouds. Just over the horizon, a southbound steamer appeared wreathe-like in and out of the snow squalls.

Driven by a northeast wind the seas were formidable, even close to the beach. They broke in a succession of four or five outward rows. The closer inshore they drove the steeper were their faces until they became concave for an instant before the faster moving tops curled over and dropped in translucent sheets to become a tumbling froth of spent energy. With the waves oblique approach to the shoreline, the breaking white water races laterally along the crests, the top edge vaporized by the wind. The spindrift peels off downwind like smoke from the stack of a fast-moving steam engine.

A large flock of Bufflehead ducks congregated just off the beach bobbing on the heaving sea. With uncanny instinct, they seem to sense which of the seas are going to break over them. An instant before they are to become immersed in the tumbling surf, they dive into the hollow face of the breaker. A second or two later they reappear on the back side of the receding wave lolling about nonchalantly, totally at home in their cold, wet world. Roger was at the Schloss doing some plumbing repair. Spying the Buffleheads he exclaimed, "It's too early for them to be around here!" Come to think about it, it always has been later in November that we have seen large flocks of these ducks passing in front of the house. They hug the shoreline around the point, skimming along just above the water.

Standing on the beach, buffeted about and chilled through by the raw wind, it takes all of five minutes to decide that there are easier ways to enjoy the roaring surf. From the vantage point of the window seat in proximity to the fire's warmth, this timeless scene can hold my attention to the total exclusion of other things going on inside. Except for the delightful smell of a goose roasting in the oven, a gift from Greg soon to be consumed with wild rice and a tangy orange sauce.

November brings nighttime temperatures which freeze the surface of the beach as hard as concrete. It's great for walking early in the morning as long as there is no wind. Instead of slogging through loose sand, it's almost like being on a sidewalk. Muggs barrels down the shore first to reach the creek where she waits excitedly to be ferried across to continue her explorations beyond. I am usually the one waiting patiently for the pokey Scottie. Sadly, I think it has something to do with age... hers not mine. She just doesn't have the old enthusiasm for sprints through the loose sand, not withstanding her awareness that the bowl at the finish line contains breakfast for which she has never lost her zest.

Fall's transformation into winter is tentative. Some of the morning's snow is on the ground, but by noon both the snow and frozen ground are gone, our surroundings reverting to the bleakness of the expiring season.

The leafless understory has exposed to view the ugly black knot infecting the smaller wild cherry trees. No matter that I've rid the property of the ulcerous manifestations of this disease each fall, there's always more to cut out and burn the next fall. These abbreviated afternoons, just before the snow sticks around for good, are a good time to get the clippers out and attack this year's crop of black knots. Cutting off the infected twigs accomplishes something and gives me a reason to be out-of-doors at a time of year when you really need a reason... rather than just doping around.

Other than in the immediate environs of the Schloss, enjoyment of the forest for the last ten days of the month is totally preempted by the ubiquitous presence of deer hunters. It would be foolhardy to walk along the footpaths in peaceful pursuits. Almost all my friends around the county are hunters and can't understand my dislike of this sport. I am well aware that uncontrolled populations of deer are not only not good for their own health, but also for the equilibrium and diversity of the entire habitat. I just don't like killing any animal; or taking a chance on maiming one and having it die an agonizingly slow and painful death, having fled the killing ground mortally wounded. My joy is seeing them alive and well, doing their thing.

I hunted birds when I was young. It seemed the thing to do in fall. When growing up I'd consumed all the tales of the wilderness and out-of-doors I could get my hands on. Book after book, all glorified the romance of the hunt and trapping and confrontations with animals, which, of course, always lost out. A genre of writing more suited to my father's time, and growing archaic by the '40s and '50s.

So the out-of-doors in fall was synonymous with hunting. Fortunately, I was not a terribly good shot so more pheasants and partridge escaped than fell to my gun. The days in the field were glorious, but each time I hit one of those beautiful birds and looked at it lifeless at my feet, I felt a terrible pang of sadness and lasting regret that I had for no other reason than enjoyment (albeit dubious in my case), taken life away from this creature which had wished only to be left alone.

My last hunt was with Tom, my constant companion in the out-of-doors, and a friend of his. That day I learned that the friend, though a nice enough person ordinarily, with a gun in his hands was an annihilator of anything that moved. The last straw was while walking back to the car at dusk. A porcupine resting high in the limbs of a dead tree was an easy target. Its grasp on the branch after having been shot was like a wrench around a nut, which slowly turns until the wrench drops off and falls to the ground with a thud. Even Tom was dismayed and embarrassed with his friend's behavior. And I was sick. But for me, at least it was a beginning. I came to realize that I enjoyed and loved animals too much to ever take another one's life. So the double-barrel shotgun went into its case in the closet never to be used again.

It seems that in this entire state almost every male is a deer hunter. Businesses close for want of employees. Boys are kept out of school to experience their rite of passage. The expressways are clogged, to say nothing of the woods, where hunters outnumber the hunted. Local restaurants and bars welcome the blip in an otherwise dead season.

After the opening weekend's barrage, the deerslayers motor home with their bragging rights lashed to fenders or splayed across the top of their cars, pleased with having done their predatory part in controlling the state's deer herd. Some of the trophies are so sad though; little yearlings with horns a couple of inches long, unsuspecting quarry, those innocent bucks. The mature ones are lying low in the swamps and other inaccessible places, if they were lucky enough to survive the initial slaughter. I wonder if the shooters of these immature animals are aware of the scorn the non-hunting world (and even some hunters) feel toward this caricature of hunting prowess.

Picture this from the next door neighbor's view. "Martha, look, Harry the Hunter's home with his deer. He shot a Bambie." Harry ought to have just stuffed the little animal out of sight in the trunk until he reached home. There he could admire his kill behind the closed doors of the garage and then, under cover of darkness, take it to a processor to have it cut up into a steak or two.

This is, in my book, more pitiable than mean-spirited. Mean-spirited is the guy who shot the blind doe raised and cared for by some friends in a small enclosed field behind their country home. This "sport" left the bright orange horse blanket in which she had been protectively wrapped for the duration of the hunting season.

\* \* \* \*

It's a pleasant coincidence that the last two weeks of November usually turn out to be tree planting time at the Schloss. (That is if Greg's superintendent can corral sufficient labor. The boss has, of course, gone hunting.) I really look forward to this event. The crew has already readied each spot by cutting out old stumps or leveling off a platform from which the hydraulic tree spade will remove the plug of earth.

Selecting trees is a wonderful way to spend a morning, walking along the periphery of stands of pine and cedars and evaluating which are likely candidates. The selection process, unfortunately, eliminates beauties that are too big. Those whose foliage on the front side looks great but are sparse on the rear because of crowding other trees become rejects, as do those which the spade can't close around without damaging other trees.

Even the color of the needles becomes a consideration. My tendency is to gravitate towards "too large" (that desire for an instant forest), so I need someone along to keep me focused on the doable. The transplant operation needs soil over six feet deep which narrows the field of choice drastically here in Door County where bedrock is frequently just under the surface. We used to get the large pines from Jim's, but there were too many tries that had to be rejected because of insufficient depth of soil. The pines in Dennis and Barbara's twenty acres have proven to be extremely hardy and recover from the shock of transplanting quickly.

The distance from where the tree is removed to the Schloss is also a consideration because of sandy soil. After the huge steel leaves of the spade close around the plug of soil containing the tree, it lifts it out of the ground and turns the plug and tree horizontal for traveling. Though the plug is tightly held by the spade, each bump in the road jars loose soil which ends up on the pavement. It's possible if the driver is in a hurry over

a bumpy road, the root system could lose all its soil. On arrival, the truck backs its rear end over the six-foot hole, turns the spade vertical and lowers the tree neatly into place. The hydraulics slip the four leaves of the spade out from around the newly planted tree and the truck is repositioned for removal of the next plug. This is dropped into the hole created by removal of the prior tree. Then the gaps in the compressed soil around the perimeter of the just planted tree plug are filled, and some fertilizer and mulch are raked over.

We have had good luck with transplanting large trees. They all seem to make it, and remarkably, a few have taken off with phenomenal growth the first summer after being transplanted, as though they never knew they were out of the ground. Transplanting at this time of year may minimize the shock. The trees are winter dormant, well beyond the end of their growing season. They have several months to settle in without need to provide immediate nourishment for new growth.

Over the years planting has become much more selective except for replacements for all the birch we're losing each year. I've long ago filled all the gaps caused by the tornado. Now it's just done when a spot catches my eye. "Hmm, a big cedar sure would look nice there." It's more likely I'll run out of money than spots because I'm always on the lookout for another to enhance a view or provide an aesthetic complement to a grouping already there. And now, of course, I can go berserk along the access lane. What is there about trees that is so intrinsically spiritual; that connects me so deeply to Mother Earth?

# Chapter 25
# Coming Home

It never occurred to me when undertaking this venture in a second home, that it would become to me an earthbound version of paradise. I use the qualification "earthbound" with trenchant intent, calling a spade a spade. Paradise in its esoteric form is a mystical la la land. This has been anything but. This individual's earthbound version of the concept has evolved while exacting a steep price. Everyone, of course, has to pay for their second home, their get-away, but in my case, at least, it's the surprises along the way that have ratcheted up the cost, both monetary and emotional. I often wonder, "What's next?" Lake Michigan might provide the answer. Though quiescent for seven years, it seems to be on another of its cyclical rises. With the basin-wide torrential rains last summer, it rose over 12 inches. The width of the beach at the point decreased ominously from 75 feet to less than ten. Fortunately fall was extremely dry, and much of the rise dropped back. But this winter has been one of heavy snow cover, and this coming fall we'll probably begin to see firsthand how good our engineering was on the shoreline armoring. The steel sheet piling and limestone have been long awaiting their ultimate battle between engineered strength and sheer mass pitted against the incredible force of storm waves.

Another hint of what's to come occurred in the fall, prior to the lake level receding. The beaching surf of storms, just ordinary storms, dammed the creek higher and higher until the floodplain did just that. The trees within the floodplain were inundated for long periods. Seven are ones that were transplanted to restore the floodplain to its appearance before the tornado and the last high-water cycle. We knew when they were planted that we were taking a chance. These cedars have thrived beyond our expectations; so much so, that getting them out of there will be impossible, without heroic measures. Resignation is now of their probable loss, not of having to move them.

To state the obvious, nothing stays the same in nature, as well as in our lives. From the standpoint of nature, this obvious fact of life hadn't touched my consciousness. Although throughout my life I have always been an interested and appreciative observer, I'd never had the opportunity, nor the good fortune, to be able to study the same piece of this planet across the seasons, year after year. Maybe I was just too wrapped up doing other things, and possibly this natural succession isn't so true in the manicured and tended lawns and gardens of suburbia. As subtly as my spirit finding its home, this microcosm of the earth's environment with all its myriad features inexorably changes and evolves before my eyes. Not being fatalistic in outlook, it's been a long pull (two grooves in the ground, a friend says) acquiring a more sympathetic understanding of this broader spectrum of natural life. The tornado wasn't the trigger. I replanted an instant forest. I wanted it to be like it was when I first saw it. It isn't, of course, but the revitalized forest is beautifully verdant… and no one but Jan or myself would recognize that such a calamity ever occurred. With the sedge grass filling in, the swamp, too, has returned to an undisturbed appearance much the same as I first saw it. On the downside, within another couple of years it's likely that not one of the larger original birch trees will remain, which is sad as these trees provide such chromatic diversity in the otherwise more somber coniferous forest. Yet, there is great satisfaction in that all the replacement birches are growing fast, and the substitute sugar maples will eventually provide a striking fall complement where there was none before.

I think my awareness of inevitable change began to dawn when, on a morning after a storm the first autumn, I found two very large cedars laying across the driveway. Their grip had undoubtedly been weakened by the tornado. It was then I began to wonder how many more of my "friends" I was going to lose as time went on. It's good there was no one to confirm what my intuition was telling me. I have learned to be more philosophical about these things, but that hasn't stopped me from helping things along. Fertilizing and spraying, planting and transplanting are just my thing, as someone else might find pleasure in propogating roses.

\* \* \* \*

Though this is paradise to me, it might not be to anyone else. Each of us has his own vision. It's like being in requited love with a woman.

There are, of course, many other attractive members of the opposite sex. It's just that love and intimacy alone make this particular woman special and unique. When it's the "real thing," the initial attraction of physical beauty and external personality deepens to one embracing soul and spirit.

As in all our loves and relationships be they person or place, I've learned there is a certain toleration and patience necessary to focus beyond the minor irritants always present. When in Alaska experiencing some of the most spectacular scenery and wildlife the world offers, I'll never forget the world-class biting flies and mosquitos. So has this piece of earth its bombardier bugs and smelly alewives, and the Schloss its mice. They are my real nemesis. We discovered them in the crawl space shortly after move-in, and they have, despite all my efforts, remained habitants of the dark (but warm) void. Contrary to most buildings of this construction type, the floor joists are not set on top of the foundation wall which might give these ubiquitous creatures avenues of entry. The floor joists in this house bear on a shelf in the foundation wall–the foundation on the outside reaching up to floor level, a continuous concrete barrier around the perimeter of the building. Mouse traps, though effective (at times I've set four traps and caught four mice in a day), just haven't cut it from the standpoint of getting rid of them once and for all time. Somewhere, somehow they have discovered or created a port of entry. Frustrated, I've spent hours on my hands and knees in the crawl space searching for their means of access. I grouted up a possible entry around the waste pipe penetration through the foundation. I did the same where the wires ran through the pipe sleeve, but to no avail.

It's become a war of attrition, trying to get them faster than they multiply. For awhile I was catching only little progeny, not the wily big ones. They're so clever and careful they sometimes manage to eat the cheese without tripping the hair-trigger trap. Containing them in the crawl space is acceptable, but every so often one manages to get "upstairs," leaving its trail across counters and beds. This is another mystery, because the only through-floor openings are for pipes and these openings were rather tightly cut, leaving little space for anything to squeeze through. It's good their invasions of the living quarters have been rare because mice in the house proper drove Jan wild, and pressure would build for a draconian solution at these times. She's not mouse scared, she just couldn't stand the implications of filth carried along and left behind. In the last year I seem to have gotten the upper hand in the unceasing battle in the crawl space, from three or four mice a weekend to one every couple of months. But their entry remains their secret. It's all part of living in the country, I guess.

The deer eat the yews and the hosta lilies, but I like the deer around as well as the flora, so I tell myself that they have to eat too. The woodpeckers and flickers occasionally hammer away at the cedar shingles getting at what's harboring underneath. There's no solution for this that I know of. Eventually, I'm going to have to replace a few shingles. Resignation is easier if the birds aren't hammering away at 5 a.m.

One day this spring with my feed-filled coffee can in hand but my thoughts elsewhere, I absently reached for the feeder... it was gone. I couldn't believe my eyes. After several years hanging by a wire pennant from a branch on the cedar tree outside the box bay, the proportedly animal-proof bird feeder had disappeared. Not only was there no feeder, nor pennant, but also no evidence whatsoever of any seeds on the ground. The twisted ends of the wire lead were difficult for me to remove with a pliers, yet for a creature working with paws. Perplexed, it took a few moments before it occurred to me to look up. There, several feet over my head, twisted upside down and hooked up to a branch by one of the perches, was the empty feeder. As the feeder had been filled to the brim the prior day, there should have been at least a pile of seed on the ground. But there was nothing. The earth was clean. I untangled the feeder, straightened the wire and refilled it—only to find it hung up in the same branch the next morning, emptied. The following weekend the trick was repeated. This wasn't doing the birds any good, and if it continued was going to rapidly deplete the bin of seed in the garage. In fact, the birds had disappeared. They couldn't even scratch a seed or two left over on the ground, so thorough was the uninvited guest. The feeder was useless. Except hoping to catch the devil, I continued to put a little feed in each day, but with always the same results.

Success apparently induced boldness. One day while fixing dinner, I caught sight of the rascal climbing the cedar with singular purpose. The raccoon climbed out on the branch, and reached downward, grabbing the pennant and then the tubular feeder (one of those Droll Yankee things) with a paw, while hanging on with two others. With incredible dexterity he raised the bottom of the feeder up to the branch until it lay horizontal on top. In a flash he'd removed the top and poured all the seed out onto the ground below. His dinner was served, and he climbed down to enjoy it–every last morsel.

Realizing that I had been bested by the raccoon, I got the stepladder out and removed the feeder with its wire pennant from the tree branch. I never returned the feeder to its branch until fall, when the raccoon had luckily moved on to new mealtime challenges.

* * * *

One of the things that has helped maintain my lifelong devotion to sailing is its aesthetics. Even during the concentration and tension of racing, the beauty of it all managed to make itself felt somewhere within my psyche. Not only is it a grand and healthy way to simultaneously exercise and relax the brain, but beyond that, sailing kept me in touch with the sea itself, being a basic facet of our natural world. The character of the sea, whether in storm or calm, is an enormous singular entity which awes us with its raw power, and conversely can tranquilize our spirit with a vista of an uninterrupted horizon at sunset or dawn. We use it as a medium to transport ourselves. We develop a relationship with it. It tolerates us. We respect it and respond to its changing moods if we know what's in our interest in staying alive, because we are so puny and vulnerable when sailing upon its vastness in small boats. Therefore, its challenges are daunting and exhilarating.

Though the sea has lost none of its magic for my spirit, I don't have to go sailing anymore to be at one with nature. Every day at the Schloss I am in touch with the natural world ashore in a relationship not requiring corollary vigilance. Only patience and appreciation are required to be connected with all the living things that comprise this particular wooded environment.

The frosting on the cake is that home is on the edge of the sea. This juxtaposition provides not only an observation post to the broad horizon, but also an opportunity for continual discovery of the interaction of the fluid sea in all its moods with the solid land. The unique rhythms of life in this narrow interactive littoral zone only became apparent with observation over time.

The Schloss in and of itself is an abode of a simple lifestyle, and the pleasures derived from it and its natural surroundings are basic and uncontrived. Just when I think I've seen it all, something new and different occurs in the scene outside. Only yesterday when arriving in the late afternoon, I was greeted with a scene out on the lake which left me breathless. From the shore-bound ice to the horizon, the lake was covered with table-flat ice of the most exquisite aqua blue. Far out there were patches of snow-covered ice, but the soft blue predominated. This mirrored surface was interrupted only by occasional fracture lines of broken ice and open water of the deepest purple, streaking jaggedly away toward the horizon.

About nine that night the ice along the shore in front of the house was suddenly and brilliantly illuminated. The tops of the glazed over cones and the myriad pieces of crystalline sea-ice piled against the shore all became individual points of fluorescent-like light. It was as though the ice had been lit from underneath. Each of the hundreds of prisms had its own scintillating geometry of intense white light. The source of this surreal ice fire was the full moon rising unseen behind the trees out at the point. Oddly, the illuminated ice was far more brilliant than was the soft glow of the light's source; on reflection a phenomenon probably caused by the refracted concentration of light through the multifaceted ice.

The stillness of the night was broken only by an occasional tonal sound of ice abrading against ice. It wasn't sharp and sudden, but resonant and musical. Its singular note, telegraphed timpanically through the medium, reached our ears with a peculiarly hollow underwater quality, like when hitting an empty steel drum with your hand, palm down.

By the following morning a wind shift had moved the vast ice field almost to the horizon, leaving open water to within a quarter mile of shore. The now hidden sun filtered weakly through a mottled overcast which moved in with the new wind. What had been an aqua blue the previous afternoon was, at this moment of time, a pale yellow contrasting boldly and abruptly with the slate gray expanse of open water offshore.

The colors of nature, whether they be as close as a flower or butterfly, or as panoramic as an evening sky evolving through a high-pressure sunset, are to me an eternal source of sensory pleasure. Oh, for the talent to be able to capture these colors on canvas! To me color is one of the most interesting characteristics of the natural world's incredible diversity. It is so purposeful an aspect in the grand design of whatever. It can be the secret of camouflage, the aphrodisia of reproduction, the herald of the seasons, a litmus of health, a projection of dominance, a reflection of something else, or simply an exhibition of beauty for beauty's sake. Color is but one enjoyable demonstration of a natural world infinite in its variety and scope, yet all inextricably and fatally linked. Connections so subtle that naturalists and scientists have only scratched the surface of awareness. We've sent men to the moon, yet are only now beginning to wake up to the dependencies so integral to the health and ultimate survival of our earth. It seems there isn't a week that passes without warning of yet another species on the brink of extinction. Are we as stewards of this natural world too late? The people and the organizations on the environmental front lines are Davids battling the Goliaths of population explosion, geo-political expediency and exploitive greed.

\* \* \* \*

There are days when it is far more pleasant to be inside looking out, and I am very content to do just that. Except, I'm forced to layer on the jackets and sweaters and thermal boots, or seasonally the foul weather gear, to walk Muggs. With advancing age, she's decided that she doesn't want to go out in this weather either, so these forays have become blessedly ever more brief and purposeful. She doesn't like getting soaked or being exposed to the bite of ten below. Indoors there's no lack of things to be done or enjoyed. Sometimes I attack my perpetual list of things to do. The indoor items always seem to remain and grow in number, while the outdoors tasks are attended to and scratched off.

An all-day fire brings a warmth not possible from the forced air furnace. The crackling flames provide a real barrier from the desolateness of the weather. Whatever Mother Nature is throwing our way, a fire holds at bay beyond the window glass any bleakness or ennui from seeping in to possess my spirit. There is a harmony within the house, and during these inclement housebound times I become more sensitized to this. I'm reawakened to the omnipresent aroma of the cedar beams and, in the bedrooms, the counterpoint of eucalyptus in its

## Coming Home

copper containers. On winter days, the sun streams in almost horizontal to bathe the rough plaster walls in shadowed patterns. It's too bright to comfortably look out onto the frozen whiteness extending to the horizon without squinting. So dominant in the interior, the massive geometry of the timber structure and the elemental nature of the stone seem to be at one with something so disparate as the grandfather clock softly chiming the quarter hours. Television noise is as incongruent to this harmony as an out-of-tune violin to a symphony. It simply doesn't fit. As the reader may suspect, I'm not a big TV viewer, nor do I need it for company.

## The Paradox of Paradise

When off duty, Muggs sighs with contentment napping in her favorite corner near her basket of treasures. Squeak toys with which she spends lengthy periods of time skulking around looking for places to hide them; Panda bear (a little worse for wear), her companion since she was no bigger than the stuffed bear; her hard rubber ball which turns her on like nothing else when she knows she is going outside to chase and retrieve; and the remains of half-gnawed chew bones. On watch, she spends her time in front of the low-silled bays or the sliding glass door watching for intruders in her domain. Her claws tick, tick, tick across the oak floor as she makes her rounds from window to doors, impressing those within hearing with her throaty browling, those self-important growls.

Threading throughout, a voice or two is the pollen of the interior bloom, carrying along the essence of sharing. Love, friendship and companionship contribute to laughter and participate in everyday nothings and solve the world's problems, if not our own. These feelings add a unique appreciation to the warmth of an interior atmosphere inherently aglow. "Come and look. There's a pileated woodpecker in the oak tree."

It has always been my regret that Pa never got to see the Schloss for it would have pleased him immensely. Door County was his favorite place on earth. Since he always loved seeing the building creations of his architect sons, the Schloss would have been a source of never-ending satisfaction. In my mind's eye, I often picture him sitting on the edge of the deck smoking his favorite pipe, soaking up the vista before his eyes.

I'm always thinking of new things to be done. Now that the garage has relieved the attic of its storage function, I've decided to finish off this space into a library with the necessary facilities for painting, drawing, writing, and alas, bill-paying. There is no more space for books, and I have dozens of these friends which I would like to have at hand. For some reason I just can't seem to simply read a book. If I like it, I've got to have it for keeps. This bibliophilic tendency poses practical space problems. This design project is at the top of the "indoor list," to have been done this winter. I even got the drawing board set up, but the unexpected got in the way of progress. One night when the outside temperature was 25 below, I battled an explosive illness so severe that irony of all ironies, the Schloss might have become my crypt. It took many weeks before I could even return to make the bed, or even think about my next project.

Throughout these weeks of absence, which seemed interminable staring at the overhead of a hospital room, I longed to be at the Schloss at home with my spirit. It became my goal. When finally I did return, I wasn't in the house five minutes when I had all the bedclothes into the washer and within an hour had thoroughly cleaned the bathroom. It was as though the demons that had possessed me when so ill were still lurking about the house. There was no way I was going to crawl into bed with that memory unexorcised. To be totally forthright, I cleaned the whole damn house!

\* \* \*

So often I've noticed that when a neighboring house is built and occupied, the new neighbors are there every spare moment. Often as time goes on, the lights are on less as interest wanes or other things become more important, driving the occupants' lives in diverging directions. Not so with the Schloss. It has been the catalyst unwittingly impelling a sea-change in my life's direction.

It all began as a convenience to facilitate working on the boat. If the aphorism, "The Dog House," had been coined by any other, it would have been "The Dog and Wife House." With its completion came the reward. It turned out far better than my expectations and hopes. The end result of this "busman's holiday" has unexpectedly given me immense creative satisfaction that I'd never expected, and it has lead me toward the realization of other dreams. Surprisingly, the needs of *Aurora* have become secondary to the enriching experiences of just being at the Schloss.

Besides existing day-to-day, is it not true that life is all about making one's dreams come true. Notwithstanding a single-mindedness of purpose from childhood on, I have been so fortunate in this regard, in this endeavor and otherwise. I achieved the wherewithal to have the luxury and the pleasure of a vacation retreat… "in a forest beside a lakeshore of grand proportion," the only place I'd ever really expressed a desire to live. Initially it provided the year-around escape from the tensions of work. Despite work always being challenging and essentially fun, now there was a counterweight on the scale of time and importance... a tug in a different direction. Each arrival at the Schloss was looked forward to with greater anticipation; each departure for the real world more difficult. The intervals away from the Schloss began to seem interminable until finally the ecstasy of escape became simply and finally, coming home... and I have arrived.

## THE END